Willett Lepley Hardin

The rise and development of the liquefaction of gases

Willett Lepley Hardin

The rise and development of the liquefaction of gases

ISBN/EAN: 9783337414276

Printed in Europe, USA, Canada, Australia, Japan

Cover: Foto ©berggeist007 / pixelio.de

More available books at **www.hansebooks.com**

THE

RISE AND DEVELOPMENT

OF THE

LIQUEFACTION OF GASES·

BY

WILLETT L. HARDIN, Ph.D.

HARRISON SENIOR FELLOW IN CHEMISTRY IN THE
UNIVERSITY OF PENNSYLVANIA

New York

THE MACMILLAN COMPANY

LONDON: MACMILLAN & CO., Ltd.

1899

PREFACE

RECENT developments in the liquefaction of air and the recent liquefaction of hydrogen have added considerable interest to the whole subject of the liquefaction of gases. The literature on this subject is scattered, for the most part, in foreign journals, and is inaccessible to a majority of those who are interested in scientific work.

The object of this little volume is to present a complete history of the development of the methods employed in the liquefaction of gases. Sufficient theory has been given to enable the popular reader to understand the principles involved. While the book has been written in a popular-science style, an effort has been made to make it of value to those who are especially interested in the subject by giving the references to the original literature.

The first intention was to include a complete account of researches at low temperatures and of the industrial applications of liquefied gases.

This, however, would make the work unduly large, and for that reason these subjects are considered only in a general way.

Full credit has been given throughout the book to the various sources of information. In conclusion I desire to express my obligations to Mr. Louis M. Thorn for the care which he has taken in the preparation of the drawings, and also to Professor Edgar F. Smith for his kindness in reading the manuscript and for many valuable suggestions.

W. L. HARDIN.

APRIL, 1899.

CONTENTS

"CONSIDER for a moment what would happen to the different substances which compose the globe, if its temperature should be suddenly changed. Suppose, for instance, that the earth . . . should be suddenly placed in a very cold region, . . . — the water which at present forms our rivers and seas, and probably a majority of the liquids which are known, would be transformed into solid mountains. On this supposition the air, or at least a part of the aëriform substances which compose it, would doubtless cease to exist in the state of an invisible fluid, for want of a sufficient degree of heat: it would return to the liquid state, and this change wou'd produce new liquids of which we have no knowledge."

LAVOISIER (1784).

LIQUEFACTION OF GASES

INTRODUCTION

PROBABLY no line of scientific research has been more productive of ingenious experiments or more prolific of results than that of the liquefaction of gases. To convert ordinary air into a liquid or even a solid is to-day a matter of little difficulty. So great an achievement, however, is not the work of a single individual or the product of a single generation. It is a result of the combined efforts of numerous experimenters, and represents the progress of a century.

The development of the methods has been accompanied by many fruitless and discouraging observations. The experiments have been of such a nature as to require the application of considerable pressure. Vessels of exceedingly brittle glass were used in most cases, and many explosions resulted. There was also a great expense attached to experiments of this nature. Within recent years various metals and alloys have been substituted for glass in the construction of apparatus. Pictet says that modern metallurgy has greatly aided in the construction of pressure-vessels. The present high grade of steel is almost indispensable.

Notwithstanding the fact that the difficulties to be overcome were enormous, and that the experimenters were subjected to considerable danger, the investigations have been carried on with untiring zeal. The history of the liquefaction of gases, like that of other lines of scientific research, is marked with periods of unusual activity and rapid progress. These periods of enthusiasm were sometimes followed by a few years of quiet, uneventful observations. The announcement of a new discovery, however, was always sufficient to give a new impetus to the work.

In tracing the development of the methods employed in the liquefaction of gases, it is perhaps advisable to divide the work into four periods. The four chapters of this volume correspond to these periods.

The first chapter is concerned with the early history of the subject. In this period some of the earlier observations on the compression of gases, and likewise a few disconnected experiments on the liquefaction of gases, are considered. These investigations are only of historical interest, and hence are only briefly outlined.

The second chapter begins with the work of

Faraday (1823), and embraces nearly half a century of fruitful observations. The fundamental methods for obtaining high pressures and low temperatures were developed to a comparatively high degree during this period. Carbonic acid, nitrous oxide, ammonia, etc., were liquefied and solidified during this time.

The third chapter is devoted to critical constants, and the continuity of the gaseous and liquid states of matter. The experiments of Andrews form the first and most important division of this chapter. These observations mark the beginning of a new epoch in the liquefaction of gases. A brief account is also given of some experiments which have been made with a view of determining the condition of matter at the critical point. The equation of Van der Waals is likewise briefly considered.

The fourth period begins with the liquefaction of the so-called permanent gases by Cailletet and Pictet (1877), and extends to the present time. During this period the apparatus employed in the liquefaction of gases has been perfected to a very high degree. The chapter is divided into four sections as follows : —

1. The pioneer experiments of Cailletet and Pictet on the liquefaction of the so-called permanent gases (1877-1882).

2. The experiments of Wroblewski, Olszewski, and Dewar from 1883 to 1895. This work may be considered as supplementary to that of Cailletet and Pictet.

3. Liquefaction of gases by the regenerative method.

4. Liquefaction of argon, hydrogen, helium, etc. The section closes with a table of physical constants.

This chapter, which extends over a period of only two decades, occupies more than one half of the present volume.

In the conclusion the three states of matter are briefly compared, and their similarities pointed out. Reference is also made to various industrial applications of liquefied gases. Great advance in this direction seems imminent from the recent progress in the liquefaction of gases. A short account is next given of physiological action, and the properties of matter at low temperatures. Here again we may look forward to great achievements. The high pressures and low temperatures which can be obtained by means of liquid air, etc., have opened up new fields of research in every branch of natural and physical science.

CHAPTER I

EARLY HISTORY

THE fact has long been known that many liquids can be converted into vapors which are similar, in most respects, to ordinary gases. It has been equally well known that the vapors thus produced can be condensed to the original liquids by lowering the temperature. These phenomena at once suggest an intimate relation between the gaseous and liquid states of matter. Some believed that the possibility of condensation applied, not only to certain vapors, but to all gases. Before the composition of the atmosphere was known, attempts were made to condense it to a liquid.

Some of the experiments which are considered in this chapter were made, not with a view of condensing the gas, but for the purpose of studying the influence of temperature and pressure on gaseous volume. The results in many cases, however, have an important bearing on the liquefaction of gases.

Experiments on the pneumatic applications of compressed air were made before the Christian

era.　These observations, and many other experiments on compressed gases, may be omitted in a
work of this nature.

For the present purpose we may begin with the
observations of Van Helmont [1] in the latter part of
the sixteenth and the early part of the seventeenth
century.　He made the first great advance in the
study of gases.　He introduced the term "gas,"
and applied it to bodies which are similar to ordinary air.　It is evident that the problem of condensing gases to the liquid state also occurred to
him, for he was the first to distinguish between
gases and vapors; the latter, he said, can be condensed to the liquid state, while the former cannot.
This distinction remained unquestioned for nearly
two centuries.　When we consider that experiments on distillations occupied fifty years of Van
Helmont's life, [2] we can understand these state-
ments, which, in reality, are far in advance of the
age in which he lived.

About the middle of the seventeenth century
Guericke [3] constructed an air-pump.　This apparatus, in its simplest form, consisted of a ver-
tical brass cylinder, which was provided with a
movable piston.　By means of this pump, he

[1] Kopp, *Geschichte der Chemie*, I, pp. 121–122.
[2] Boerhaave, *Elements of Chemistry*, Eng. ed., I, p. 16.
[3] Haüy, *Nat. Philos.*, Eng. ed., p. 220.

studied the influence of pressure on the volume of a given quantity of air. The principle of Guericke's air-pump has been applied by numerous experimenters in their study of compressed gases.

The first systematic observations on the influence of pressure on the volume of a gas are those of Boyle. In 1662 he announced an important law, which bears his name. He stated that the volume of a gas varies inversely as the pressure. In other words, the product of the pressure and volume is a constant. The law is usually expressed by the equation

$$pv = c,$$

where c is a constant. The same law was announced by Mariotte a few years later. The conditions under which gases deviate from this law will be considered in a subsequent chapter.

From the writings of Boerhaave (1731), it is evident that he had considered the problem of liquefying and of solidifying ordinary air. In the English edition of his *Elements of Chemistry*, pp. 249, 250, we find the following statement: —

" The first property, then, of air which offers itself to our consideration is its fluidity. This is so natural to it that I do not remember ever to have heard of any experiment by which air could be deprived of it. It is evident to every one's

observation that even in the sharpest frost, when
almost everything is congealed, the air still remains fluid ; nay, in an artificial cold, 40° greater
than nature has ever been known to produce, the
air still retained its fluidity. . . . If you compress
the air, with ever so great weight and force, into
the utmost density, it does not then become solid
by concretion, but remains equally fluid as before.
. . . I have never yet met with a single experiment
by which it appeared that air was coagulated into
a solid mass. I confess that, one noon in frosty
weather, when the air was very serene, I observed
some very small corpuscles floating about in it. . . .
But, after a careful observation, I discovered that
these were nothing but little globules of water,
which were congealed, and which appeared in the
form of a very subtle hoar-frost."

In speaking of the elasticity of the air, the same
writer, pp. 263–264, says : —

" Another law which we find to hold true is, that
the elasticity of the air cannot be destroyed. . . .
In whatsoever manner the air has been compressed by the utmost power of weights, it has
always remained very fluid ; for, after it has been
contracted into the greatest density, it has constantly restored itself again into all its particles,
so as to fill up exactly the former space ; all particles retreating with the same ease with which they

came together. . . . We may fairly assert that the fluidity of the air in all the large compass, from the most rarefied to the most compressed, remains without alteration ; and that therefore it is neither capable of being solidified by the intensest cold, nor the greatest degree of compression."

On page 266 of the same book we find that Boerhaave was familiar with the fact that heat expands the air. The following are his words :—

" By the application of fire the air becomes so rare that neither the measure nor limit of its dila-tion has yet been discovered. . . . Air of unequal masses, but of the same density, is always ex-panded in the same measure by the same degree of fire: so that these expansions in the same density of air are, by a constant law of nature, always proportional to the augmentations of heat."

As early as 1702 Amontons[1] studied the effect of temperature on the elastic force of the air. He says: "If equal or unequal masses of air are charged with equal weights, their elastic forces will be equally increased by equal degrees of heat."

In 1787 Charles[2] called attention to the fact

[1] Haüy, *Nat. Philos.*, Eng. ed., pp. 255–260.
[2] Charles communicated his results to Gay Lussac and did not publish them.

that all gases expand equally with the same increase in temperature. Dalton[1] made similar observations in 1801. He is also the author of the following statement:[2] "There can scarcely be a doubt entertained respecting the reducibility of all elastic fluids of whatever kind into liquids; and we ought not to despair of effecting it in low temperatures, and by strong pressures exerted upon the unmixed gases." Dalton, just as Lavoisier, foresaw the result of subjecting gases to intense cold.

The relation of temperature to the volume of a gas was more thoroughly investigated by Gay Lussac in 1802. From his results, as well as from those of Dalton and Charles, we find that an increase of 1° in temperature[3] increases the volume of a gas by about $\frac{1}{273}$ of the volume at 0°, provided the pressure remains constant. In other words, the volume v of a gas at the temperature t is given by the equation,

$$v = v_0\left(1 + \frac{t}{273}\right),$$

where v_0 represents the volume at 0°. In various

---

[1] *Manchester Memoirs*, V, p. 535, 1801. Ref. Roscoe and Schorlemmer, *Chemistry*, I, p. 60, 1895.

[2] Ref. Dewar, *Proc. Roy. Inst.*, 8, p. 657, 1878.

[3] All temperatures will be given in Centigrade degrees unless otherwise expressed.

text-books, this law is referred to as the law of Charles, the law of Dalton, or the law of Gay Lussac, depending upon the author of the book. According to this law all gases have the same coefficient of expansion with regard to temperature. In this respect the gaseous state of matter stands alone; the coefficients of expansion of liquids and solids show considerable variation.

Absolute Temperature.

Throughout the discussion of the experiments on the liquefaction of gases, reference will frequently be made to absolute temperature. The significance of this term may be best understood by considering the law of Gay Lussac. Suppose we start with a given volume of gas at 0°. Let the pressure remain constant. If the temperature be increased to 273°, the volume will be doubled. If, on the other hand, the temperature be lowered, the volume will be decreased by $\frac{1}{273}$ of the original volume for each degree of temperature change. If the law remains valid for all temperatures, an important conclusion follows; namely, at the temperature of $-273°$ the volume of the gas will become equal to zero. Of course, it is not likely that such a condition will ever be attained, for all gases, so far as

known, deviate from this law before the temperature of $-273°$ is reached. The temperature
at which the volume of a perfect gas becomes
equal to zero ($-273°$ C. or $-460°$ F.) is called
the ABSOLUTE ZERO OF TEMPERATURE. Absolute
temperatures are measured from this point. The
absolute zero may also be defined as that temperature at which the kinetic energy of the
molecules is equal to zero. Thomson (now Lord
Kelvin) has also proposed a definition which is
based entirely upon energy considerations.[1] This
definition, however, need not be discussed here.
He obtained the value $-273°.7$ for the absolute
zero.

Both Dalton and Gay Lussac made observations on the change of temperature which accompanies the expansion or compression of a
gas. Dalton[2] found that "about $50°$ of heat
are evolved when air is compressed to one half
of its original volume, and that, on the other
hand, $50°$ are absorbed by the corresponding
rarefaction." Gay Lussac[3] employed two large
flasks, one of which contained the gas, while the

---

[1] *Phil. Mag.*, Oct., 1848.

[2] *Memoirs of the Literary and Philosophical Society of Manchester*, V, Part II, pp. 251–525.

[3] See original memoir in *The Free' Expansion of Gases*, by
Ames, pp. 3–13.

other was exhausted. Thermometers were placed in each of these flasks, and the two were then connected. This allowed the gas to expand in one flask, while it was being compressed in the other. In this manner, Gay Lussac experimented with ordinary air, hydrogen, carbon dioxide, and oxygen, and observed the accompanying changes of temperature. This subject has been more thoroughly investigated by Joule and Thomson. Their work will be referred to in a subsequent chapter. This method, in recent years, has become of great importance in the production of low temperatures.

It has been considered probable by some that Count Rumford,[1] in his experiments to determine the explosive force of gunpowder, may have liquefied carbonic acid gas. He exploded the powder in cylinders closed with a weighted, movable valve. The following are his words: "When the force of the generated elastic vapor was sufficient to raise the weight, the explosion was attended by a very sharp and surprisingly loud report; but when the weight was not raised, as also when it was only a little moved, — but not sufficiently for the elastic vapor to make its escape, — the report was scarcely audible at the

[1] *Phil. Trans.*, 1797, p. 222. See also *Alembic Club Reprints*, No. 12, p. 20.

distance of a few paces. . . . In many of the experiments, in which the elastic vapor was confined, this feeble report attending the explosion of the powder was immediately followed by another noise totally different from it, which appeared to be occasioned by the falling back of the weight." He also calls attention to the small degree of expansive force of the confined elastic vapor when it was allowed to escape. Instead of rushing out with a loud report, the gas escaped with a hissing noise. The aqueous vapor produced in these experiments was evidently condensed to a liquid, but, in the light of the researches of Andrews, it seems that the temperature produced by the explosion would be higher than 31°, — the critical temperature of carbon dioxide. If this is true, the carbonic acid, of course, would not have been condensed to the liquid state.

In 1799 Van Marum [1] published an account of some experiments which were performed several years earlier. The main object of these observations was to study the effect of low pressures on different liquids. Some of the liquids, he stated, were entirely converted to vapors by this method. When the pressure was increased, these vapors

---

[1] *Gilbert's Ann.*, I, p. 145.

condensed again to the liquid state. By means of an air-pump, he compressed ammonia gas, and found that, as the pressure increased, the volume did not decrease in accordance with Boyle's law. At a pressure of three atmospheres, he says, drops of liquid ammonia were formed. It is very probable, however, that the gas used in this experiment contained a small quantity of aqueous vapor.

Fourcroy and Vanquelin[1] made a series of experiments, in which they subjected certain gases to low temperatures. In Accum's Chemistry, I, p. 337, we find the statement that ammonia gas was liquefied in these experiments. It seems, however, that aqueous ammonia was used in these observations, and that the pure gas was not liquefied. These experimenters also subjected hydrochloric acid gas, sulphuretted hydrogen, and sulphurous acid gas to low temperatures. They did not succeed, however, in liquefying any of these gases. The authors state that they obtained a temperature of − 40°.

During this same year Guyton de Morveau[2] experimented with ammonia gas at low temperatures. It is probable here, as in the observations of Fourcroy and Vanquelin, that the gas contained moisture. The method of drying was to pass the

[1] *Ann. de Chim.*, 29, p. 281.
[2] *Ann. de Chim.*, 29, pp. 290, 297.

gas into a glass vessel, the temperature of which
was − 21°.25. The object of this was to convert
the aqueous vapor into ice. From this balloon
the uncondensed gas passed to a second glass
vessel, where it was subjected to a temperature
of − 43°.25. At this temperature drops of liquid
formed on the interior surfaces of the containing
vessel. The author states that ammonia, dried
as completely as possible by subjecting to a temperature of − 21°, condenses to a liquid at − 48°.
It is not likely that the gas was rendered completely dry by this method; and the author calls
attention to the fact that the liquid, produced
at − 48°, probably contained a small quantity of
water.

In 1805 Stromeyer subjected arsine to low temperatures. Thenard[1] says that the gas was condensed to a liquid in these experiments. Another
reference[2] states that "Professor Stromeyer condensed the gas so far as to reduce it in part to a
liquid, by immersing it in a mixture of snow and
muriate of lime, in which several pounds of quick-
silver had been frozen in the course of a few
minutes." Faraday remarks that, "from the circumstance of its being reduced only in part to a
liquid, we may be led to suspect that it was rather

---

[1] *Traité de Chimie*, I, p 373.
[2] *Nicholson's Jour.*, 19, p. 382.

the moisture of the gas that was condensed than the gas itself."

Numerous text-books refer to the liquefaction of sulphurous acid gas by Monge and Clouet. Thomson, in his System of Chemistry, 6th ed., II, p. 118, says the gas was condensed when exposed to a temperature of − 18°. In Accum's Chemistry, I, p. 345, and Murray's Chemistry, II, p. 405, we find the statement that the condensation of sulphurous acid was effected by the application of strong pressure and intense cold.

The experiments of Northmore[1] in 1805–1806 are among the most important of the earlier observations. His object was to determine the effect of pressure on the affinities of gases. His results, however, show that some of the gases were liquefied. The apparatus consisted of an exhausting syringe, a condensing pump, a connecting spring-valve, and glass receivers. When chlorine was compressed, he observed the formation of a yellow, extremely volatile, fluid. This appears to be the first reliable account of the liquefaction of chlorine. He also liquefied hydrochloric acid gas and sulphurous acid gas. The latter experiment, he says, corroborates the statement of Monge and Clouet, that sulphurous acid is condensed to a

[1] *Nicholson's Jour.*, 12. p. 368, and 13, p. 233. Also *Alembic Club Reprints*, No. 12, pp. 69–79.

c

liquid by the simultaneous application of strong pressure and low temperature. In the experiments on carbonic acid, Northmore remarks that the receiver very unexpectedly bursts with violence.

These observations were very successful, considering the period in which they were made. It is difficult to understand why the results obtained by Northmore were no incentive to further investigation. The fact remains, however, and these experiments have only an historical interest.

In 1822, Cagnaird de la Tour[1] made a rather extended series of experiments by heating liquids in sealed glass tubes. This work is of considerable importance, in as much as it is a forerunner of the researches of Andrews on critical constants. Tour called attention to the following facts : —

1. Alcohol, naphtha, and ether, when subjected to heat and pressure, are converted into vapors in a space slightly greater than double that of each liquid.

2. The increased pressure, due to the presence of air in the tube, did not prevent the evaporation of the liquid in the same space. The expansion, however, was more regular.

3. The experiments on water were unsatisfactory, owing to imperfections in the apparatus.

---

[1] *Ann. de Chim. et de Phys.*, 21, pp. 127, 178.

Ether is completely vaporized in a space somewhat less than double that of the liquid, at a temperature of about 160°. The pressure exerted by
ether at this temperature is from 37 to 38 atmospheres. At a temperature of 200°, alcohol vaporizes
in a space less than three times that of the liquid.
The pressure in this case is about 119 atmospheres.

The measurements on ether are given in the
following tables :[1] —

Volume of Liquid = 7		Volume of Vapor = 20
Temperature in Reaumur Degrees	Pressures in Atmos- pheres	Differences
80°	5	—
90	7	2
100	10	3
110	12	2
120	18	6
130	22	4
140	28	6
State of Vapor 150	37	9
160	48	11
170	59	11
180	68	9
190	78	10
200	86	8
—	—	—
—	—	—
—	—	—
260	130	—

[1] *Ann. de Chim. et de Phys.*, 22, pp. 411, 412.

The results obtained by heating one half of this quantity of liquid in a tube of the same volume were as follows: —

Volume of Liquid = 3½		Volume of Vapor = 20
Temperature in Reaumur Degrees	Pressures in Atmospheres	Differences
100°	14	—
110	17	3
120	22	5
130	28	6
140	35	7
State of Vapor 150	42	7
160	50	8
170	58	8
180	63	5
190	66	3
200	70	4
—	—	4
—	—	4
—	—	3
260	94	4

The important part of these observations is, that the liquid disappeared at the same temperature in both cases. This was an indication that, regardless of the pressure, the substance cannot exist in the liquid state when heated above a certain temperature. These results exerted but little influence until after the work of Andrews nearly fifty years later.

So far the investigations are only of historical interest. The results which are recorded on the preceding pages have been of little value to the later experimenters. From this point the work becomes more systematic. The interesting observations of Cagnaird de la Tour may be said to close the first chapter in the liquefaction of gases.

CHAPTER II

HERE, as in many other lines of scientific research, we must turn to Faraday for the first systematic observations. This great experimenter has taken the initiative step in many lines of scientific investigation, and is the author of numerous and important discoveries. In 1823, Faraday accidentally liquefied chlorine. Although an accident, he perceived the full significance of the result, and followed the observation with two elaborate series of experiments, in which he exhausted all his resources in the endeavor to liquefy the so-called permanent gases.

At the suggestion of Davy, Faraday experimented with the hydrate of chlorine and studied the effect of heating it in a closed glass tube. When placed in water at 60°, he observed no change; but when placed in water at 100°, a yellow gas was set free. On cooling, this gas condensed to a yellow liquid which resembled the chloride of nitrogen. During the progress of this experiment, Dr. Paris[1] happened to enter the lab-

---

[1] Tyndall, *Faraday as a Discoverer*, p. 14.

oratory. " Seeing the oily liquid in the tube, he rallied the young chemist for his carelessness in employing soiled vessels. . . . Early the next morning, Dr. Paris received the following note : —

" ' DEAR SIR, — The *oil* you noticed yesterday turns out to be liquid chlorine.

" ' Yours faithfully,

" ' M. FARADAY.' "

The chlorine had been liquefied by pressure. When the tube was opened, the contents exploded.

Faraday made a number of observations to prove that this yellow liquid was chlorine. Among other experiments, he subjected dry chlorine gas to pressure, and obtained a liquid which was similar in all respects to that obtained from the hydrate. He also made some approximate determinations of the density of liquid chlorine.

Almost simultaneous with this work of Faraday, Davy[1] liquefied hydrochloric acid gas by treating ammonium chloride with sulphuric acid in a closed tube. He then substituted ammonium carbonate for the chloride, and endeavored to liquefy carbon dioxide. Only one experiment was

[1] Note to Faraday's article on " Liquefaction of Chlorine."

made, and, in that case, the tube burst. At the
suggestion of Davy, Faraday continued the work
alone.

In these experiments, a number of gases were
subjected to pressure, and the results were published in one article[1] in 1823.

Sulphurous Acid

Mercury and concentrated sulphuric acid were
placed in a closed, bent glass tube which was
afterward sealed (Fig. 1). The end of the tube
containing the reacting
substances was heated,
while the other end was
kept cool by means of
moistened paper. Sul-
phurous acid gas was gen-
erated, and, after the
sulphuric acid had become
saturated, it was evolved, and condensed to a liquid
in the cold end of the tube. Faraday remarks
that sulphurous acid forms a limpid, colorless,
highly fluid liquid, which does not solidify at 0° F.
When the tube was opened, a portion of the liquid
evaporated rapidly. This lowered the tempera-

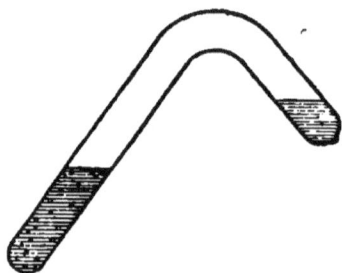

FIG. 1.

[1] *Phil. Trans.*, 113, pp. 189–198. See also *Alembic Club
Reprints*, No. 12, pp. 10–19.

ture of the remaining portion, so that it remained a liquid for some time at the ordinary
atmospheric pressure. A piece of ice thrown
into the liquid caused rapid boiling. Faraday
then took dry sulphurous acid gas, and subjected
it to pressure in a glass tube by means of a
syringe. When the tube was cooled to 0° F., the
gas condensed to a liquid which was similar to
that obtained from the mercury and sulphuric acid.

Sulphuretted Hydrogen

In this experiment the sealed glass tube contained hydrochloric acid and sulphuret of iron.
On warming the contents, sulphuretted hydrogen
was generated and condensed in the longer end
of the tube, which was surrounded by a mixture of
ice and salt. The resulting liquid was colorless,
limpid, and excessively fluid. By the introduction
of a small gauge into the tube, he found the pressure to be nearly 17 atmospheres at the temperature of 50° F.

Carbonic Acid

The materials used in this case were ammonium
carbonate and concentrated sulphuric acid. The
glass tubes, however, were much stronger than
those used in the previous experiments. Even
then a number of violent explosions resulted.

The carbonic acid was condensed to a color-less, extremely fluid liquid. The vapor pressure at 32° F. was measured, and found to be 36 atmospheres.

Euchlorine

This gas was generated from a mixture of potassium chlorate and sulphuric acid in a closed tube. On cooling, it condensed to a yellow, transparent liquid.

Nitrous Oxide

Ammonium nitrate, dried by heating in the air to partial decomposition, was heated in a closed bent tube. When the gases were cooled, two liquids separated. One was found to be water containing a little nitrous oxide in solution, while the other was nitrous oxide. The appearance of the liquid was similar to that of carbonic acid. The vapor pressure at 45° F. was found to be 50 atmospheres. These experiments were accompanied by a number of violent explosions.

Cyanogen

Pure, dry, mercuric cyanide was treated in a manner similar to that of ammonium nitrate. The resulting cyanogen was condensed in the cold end of the tube to a limpid, colorless, liquid.

The vapor pressure at 45° F. was found to be 3.6 atmospheres.

Ammonia

Dry silver chloride was allowed to absorb pure, dry ammonia gas. The substance was then heated in a closed tube. The gas was again set free, and was condensed to a colorless, transparent liquid in the longer arm of the tube, which was cooled by means of ice and water. The vapor pressure at 50° F. was about 6.5 atmospheres.

Faraday repeated the experiments of Davy on hydrochloric acid. He also determined a number of physical constants of these various liquids. He attempted to liquefy hydrogen, oxygen, etc., but without success. Twenty-one years later, he published another series of experiments on the liquefaction of gases. These observations will be considered in their proper place. Meanwhile we may turn our attention to the work of other investigators.

EXPERIMENTS OF PERKINS

In a note,[1] published in 1823, Perkins calls attention to a series of experiments on compressed air. The details of the work were not published until a few years later. The primary object of the

[1] *Annals of Phil.*, VI, p. 66.

investigation was to study the effect of high pressure on liquids. The author states, however, that
when ordinary air was subjected to a pressure of
1000 atmospheres, drops of liquid began to form,
and that at 1200 atmospheres, a beautiful transparent liquid could be seen on the surface of the
mercury. The liquid remained in the tube after
the pressure had been removed. Perkins supposed that the liquid thus formed was condensed
air. From the researches of later experimenters,
however, it is evident that ordinary air cannot be
liquefied under such conditions. The liquid which
appeared at the surface of the mercury was evidently formed by the condensation of aqueous
vapor. Perkins also remarks that he liquefied car-
buretted hydrogen at a pressure of 1200 atmospheres. He developed a method for obtaining
extremely high pressures.

Experiments of Bussy

In 1824, Bussy[1] published a series of observations on the liquefaction of gases. His method of
procedure was entirely different from that of Faraday. He subjected the gases to low temperatures,
but did not increase the pressure. He condensed

[1] *Ann. de Chim. et de Phys.*, XXVI, p. 63; *Pogg. Ann.*, I, p. 237,
1824.

sulphurous acid to a colorless, transparent liquid, at the ordinary atmospheric pressure, by subjecting it to a temperature of -18 to $-20°$. Bussy observed that if the liquid sulphurous acid was allowed to evaporate in the air, the temperature sank to $-57°$, and that if evaporated under reduced pressure, the temperature sank to $-65°$. His work was a great advance in the liquefaction of gases, inasmuch as he made use of this method to produce low temperatures. By means of liquid sulphurous acid, he liquefied chlorine and ammonia, and in a similar manner he liquefied and solidified cyanogen. This method of lowering the temperature soon came into general use for both scientific and industrial purposes.

Heat of Vaporization

Inasmuch as the evaporation of liquids becomes, at this period, an important method for the production of low temperatures, we may briefly consider, at this point, the thermal change which accompanies the passage of a substance from the liquid to the gaseous state.

When a liquid changes to the gaseous state, a certain quantity of heat is absorbed. The heat required to change one gram of a liquid to a vapor, at the same temperature and pressure, is called the *heat of vaporization*. The heat absorbed is used : —

1. To increase the internal energy of the substance; *i.e.* the energy of the particles must be sufficiently increased to overcome the force of cohesion.

2. To perform external work; *i.e.* the volume of the substance must be increased against a definite pressure.

Let E represent the energy of the vapor, and E_1 the energy of the liquid, then the increase in the internal energy is

$$E - E_1.$$

If V and V_1 represent the volumes of the vapor and liquid, and p is the pressure, which remains constant, then the external work performed is equal to

$$p\,(V - V_1).$$

The total energy and work required to vaporize a liquid, therefore, is

$$E - E_1 + p\,(V - V_1).$$

Knowing the mechanical equivalent of heat, this expression may be represented in thermal units; in this latter form it represents the heat of vaporization. The numerical value of the heat of vaporization depends upon the temperature.

Inasmuch as heat is absorbed during evaporation,

it is evident that a liquid cannot maintain a constant temperature during the process of evaporation, unless it receives heat from some external source. Ordinarily, there is a constant flow of heat from the surrounding bodies to the liquid during evaporation.

Suppose, however, that this supply of heat from the surrounding bodies is cut off by insulation, then the heat required in the evaporation must be drawn from the liquid itself. The temperature, of course, will be lowered by this process. If the evaporation be hastened, e.g. by reducing the pressure, the temperature will be reduced more rapidly and to a much greater degree. Leslie has shown that water can be frozen by means of its own evaporation under reduced pressure.

EXPERIMENTS OF COLLADON

In 1828 Colladon [1] constructed an apparatus for the purpose of liquefying ordinary air. The general plan of the apparatus is shown in figure 2.

The pressure was produced by means of an hydraulic pump, and transmitted through the tube Cc to the interior of a strong steel cylinder B, which was partially filled with mercury. In this cylinder was placed a glass tube T which was

[1] Pictet, *Ann. de Chim. et de Phys.* [5], XIII, p. 226.

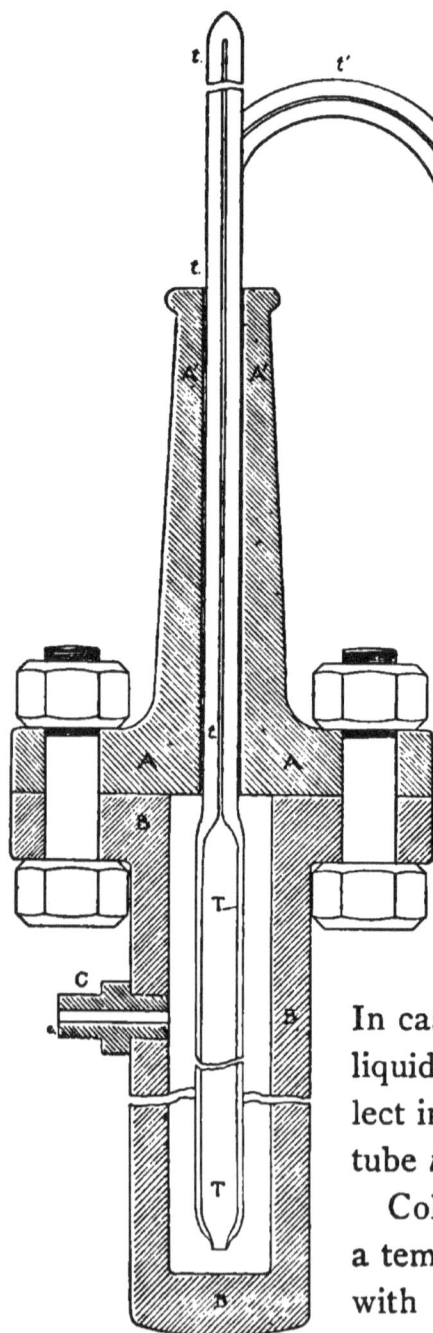

FIG. 2.

open at the lower end, but fused at the upper end to the thick-walled, closed glass tube t, the interior diameter of which was from 1.5 to 2 mm. This tube t projected from the cylinder through the elongated cover A, above which it was bent downward to t' as shown in the figure. The bent-down portion of the tube was placed in a freezing mixture. In case of condensation the liquid, of course, would collect in the lower end of the tube t'.

Colladon experimented at a temperature of $-30°$ and with pressures as high as

400 atmospheres.   The results, however, were all
negative.

EXPERIMENTS OF THILORIER

In 1834 Thilorier [1] liquefied carbonic acid on a
much larger scale than had hitherto been done.

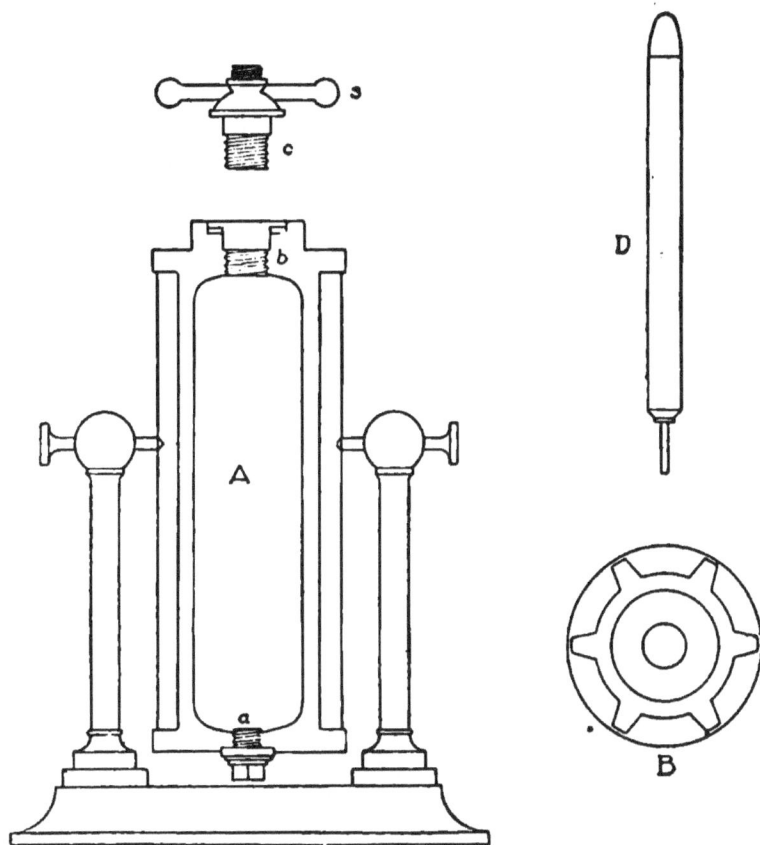

FIG. 3.

The general plan of the apparatus [2] employed is
shown in figures 3 and 4.   The gas is generated

[1] *L'Institut*, 58, p. 197; *Ann. de Chim. et ae Phys.*, 60, pp. 427, 432.
[2] *Lieb. Ann.*, 30, p. 122.

D

in the wrought-iron vessel *A*, and compressed by means of its own pressure in the wrought-iron receiver *F*. These two vessels are similarly constructed. *B* represents a cross-section of the cylinder *A*. This vessel is supported by means of two pivots so that it can be rotated. Eighteen hundred grams of bicarbonate of sodium, and $4\frac{1}{2}$ litres of water are placed in the generator. The vessel *D* is then filled with concentrated sulphuric acid, and placed in the cylinder with the sodium bicarbonate and water. The stopper *c* is then screwed tightly into the top of the generator, and connection made with the receiver by means of the copper tube *E*. On rotating the vessel *A*, the sulphuric acid in *D* comes in contact with the carbonate. The carbonic acid gas which is evolved in the reaction produces a very high pressure in the generator. After several minutes the stop-cocks *mm* are opened. The gas rushes immediately into the receiver until the pressures in the two vessels are equal. A portion of the gas is liquefied by pressure. The quantity of liquid produced may be largely increased by surrounding the receiver with a freezing mixture. When equilibrium has been established between the two vessels *A* and *F*, the stop-cocks *m* and *m* are closed. The pressure in the generator is then relieved, the contents are removed, and a new charge is introduced. The process is repeated in this manner until from 2 to 3 litres of liquid carbonic acid have

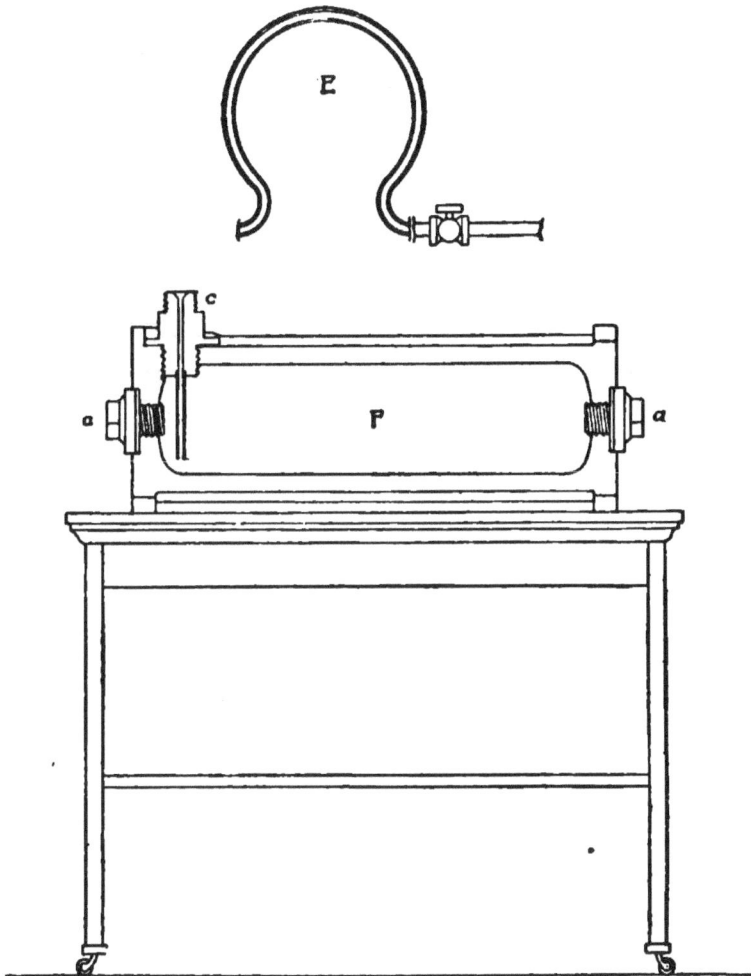

FIG. 4.

collected in the receiver. From five to ten repetitions are usually required.

In this way, Thilorier obtained liquid carbonic

acid in rather large quantities, and made a series
of observations on its physical properties.   He
measured the vapor pressure at different temperatures and found it to be 36 atmospheres at 0°,
and 73 atmospheres at 30°.   He also determined
the specific gravity, and made observations on the
thermoscopic effect of the liquid.   The condensed
gas was found to be insoluble in water and fat
oils, but soluble in all proportions in alcohol, ether,
naphtha, turpentine, and carbon disulphide.   It
reacts with metallic potassium, but produces no
effect on lead, tin, copper, iron, etc.   The abnormal expansion of the liquid carbonic acid was also
investigated.

Thilorier made some important observations on
the production of low temperatures by the evaporation of liquid carbonic acid.   When a jet of the
liquid was directed upon the bulb of an alcohol thermometer, the temperature sank rapidly to −90°.
The sphere of action, however, in this as well as
in other cases was limited almost entirely to the
point of contact.   The congelation of mercury was
confined to small portions of it; and when the
hand was exposed to a jet of the liquid, a burning
sensation was felt, but the effect was confined to
very narrow limits.

The author explained this limited sphere of
action by the low conductivity and small capacity

for heat of carbonic acid. He says: "If gases have little effect in the production of cold, it is not so with vapors whose conductivity and capacity for heat are much greater. I have, therefore, thought that if a permanent liquid — ether, for example — could be placed under the same conditions of expansibility as liquefied gases, we might obtain a frigorific effect much greater than that produced by liquid carbonic acid. To accomplish this, ether must be rendered *explosible*, and this I have easily effected by mixing it with liquid carbonic acid. In this intimate combination of the two liquids, which dissolve each other in all proportions, ether ceases to be a permanent liquid at the ordinary atmospheric pressure; it becomes expansible like a condensed gas, still preserving its properties of a vapor; namely, its conductivity and capacity for heat."

The effects of a jet of explosible ether are more pronounced and extend over a much greater area than those produced by a jet of liquid carbonic acid. Fifty grams of mercury can be solidified in a few seconds by means of this mixture.

Solidification of Carbonic Acid

Thilorier not only liquefied carbonic acid, but succeeded in converting it into a solid. The

receiver which contained the liquid carbonic acid
was connected with a vessel formed of two separate halves (Fig. 5) by means of a small tube.
The sudden expansion produced a very low temperature, and a portion of the gas solidified. The
author thought, however, that the solid particles
were composed of ordinary ice which resulted from
the freezing of aqueous vapor in the air. The
experiment was repeated before Arago, Thenard,

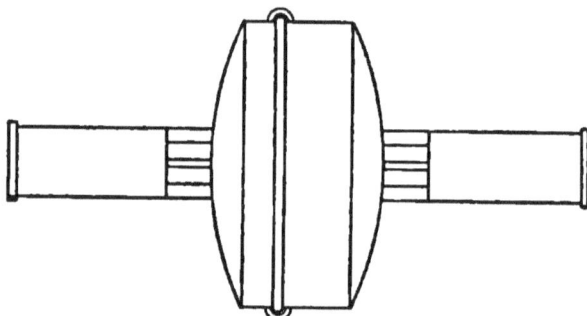

FIG. 5.

and Dulong. The solid portions were then found
to be carbonic acid. "Gaseous at the ordinary
temperature and pressure, and liquid at 0° under a
pressure of 36 atmospheres, carbonic acid becomes
solid at a temperature of about 100° below zero,
and retains this new condition for several minutes
in the open air, without the necessity of any compression."

The author calls attention also to the remarkable
difference in the behavior of the liquid and solid

acid.   While in the liquid state, he says, the
elastic force is exceedingly great, — being equal
in explosive power to an equal weight of gunpowder.   The solid, on the other hand, disappears
insensibly by slow evaporation.   A fragment of
solid carbonic acid, he adds, slightly touched by
the finger, glides rapidly over a polished surface,
as if sustained by the gaseous atmosphere with
which it is constantly surrounded, until it is entirely
dissipated.   The vaporization of solid carbonic
acid is complete.   It leaves but rarely a slight
humidity, which may be attributed to the action of
the air on a cold body, the temperature of which
is far below that of freezing mercury.

The solidification of large quantities of carbonic
acid has been of great value to later experimenters
for the production of low temperatures.   The solid
acid moistened with ordinary ether is known as
Thilorier's Mixture.   A temperature of

$$-110° = -166° \text{ F.}$$

can be obtained by means of this mixture.

Modification of Thilorier's Apparatus

In a paper read before the British Association
for the Advancement of Science,[1] Addams calls
attention to three forms of apparatus which had

[1] *Proceedings* for 1838, p. 70.

been used by him to liquefy and solidify carbonic acid on a large scale. The first method is mechanical, in which powerful hydraulic pumps are used to force the gas from one vessel into a second, by filling the first with water, saline solutions, oil, or mercury. The second form of apparatus is a modification of that invented and used by Thilorier. The third includes the mechanical and the chemical methods, by means of which much of the acid formed in the generator is preserved; whereas by the arrangement of Thilorier, two parts in three rush into the atmosphere and are lost. The apparatus is provided with two gauges, — one to determine when the generator is filled with water, and the other to indicate the quantity of liquid acid in the receiver. By means of this apparatus, Addams prepared liquid carbonic acid in considerable quantity and measured the elastic force at different temperatures. The solid carbonic acid used by Faraday in his second series of experiments on the liquefaction of gases was prepared by Addams.

EXPERIMENTS ON HYDROGEN AND OXYGEN

In these experiments Maugham[1] attempted to liquefy hydrogen and oxygen by means of press-

---

[1] *Proc. Brit. Assoc. for Adv. of Sci.* (1838), p. 73.

ure. The pressure was obtained by the electrolysis of acidulated water in closed, strong glass U-tubes. In this manner, the author was able to burst tubes of the strongest glass. He suggested that if the experiments were carried out at low temperatures, the hydrogen and oxygen would probably condense to the liquid state. This method for obtaining high pressures has been used by some of the later experimenters.

OBSERVATIONS OF TORREY

These experiments were not published by the author, but the results were communicated in private letters to Dr. Silliman, who published them two years later.[1] The editors make the following statement: —

" We have been, from time to time, informed by letters from Prof. John Torrey of New York, of his progress in the condensation of gases, and although not intended for publication, we now take the liberty to give some citations from his letters."

April 11, 1837, Torrey writes that, although he was unable to procure a suitable form of apparatus in which to condense carbonic acid, he had successfully carried out numerous experiments

---

[1] *Silliman's Jour.*, 35, p. 374, 1839.

with glass tubes. A month later he forwarded a tube containing liquid carbonic acid to Dr. Silliman. Torrey also liquefied sulphurous acid and chlorochromic acid. Perfectly dry phosphorus, he says, is not inflamed by liquid chlorochromic acid, but if moist in the slightest degree, it will burn with a loud explosion, requiring particular precautions. In another letter, he remarks, "I have been shooting with an air-gun, using liquid carbonic acid for throwing the balls, and I hope soon to emulate Perkins' steam gun."

EXPERIMENTS OF MITCHELL

In 1839 Mitchell[1] modified the apparatus of Thilorier somewhat, and obtained both liquid and solid carbonic acid. When the liquefied gas is allowed to escape into an iron receiver, he says, "a large portion of the liquid is instantly expanded into gas which escapes through a tube. The coldness consequent on the enormous expansion freezes another portion of the liquid, which then falls to the bottom of the receiver. About one drachm of solid matter is thus formed for each ounce of liquid."

Mitchell also made a series of observations on the physical constants of liquid and solid carbonic

---

[1] *Silliman's Jour.*, 35, p. 346.

acid. By means of the solid acid he solidified mercury, and experimented with the solid metal. Liquid sulphurous acid freezes at about $-80°$, while ordinary ether, he says, when subjected to a temperature of $-100°$ is not, in the slightest degree, altered. He further adds that, "when a piece of solid carbonic acid is pressed against a living animal surface, it drives off the circulating fluids, and produces a ghastly white spot. If held for fifteen seconds it raises a blister, and if the application be continued for two minutes, a deep white depression with an elevated margin is perceived; the part is killed, and a slough is in time the consequence. I have thus produced both blisters and sloughs, by means nearly as prompt as fire, but much less alarming to my patients." Experiments were also made with a number of liquids and metals, by placing them in liquid carbonic acid.

EXPERIMENTS OF AIMÉ (1843)[1]

In these experiments, ethylene, nitric oxide, nitrogen, hydrogen, carbon monoxide, and oxygen were subjected to very high pressures. The pressure was obtained by sinking the gases in suitable vessels to great depths in the ocean. The method,

[1] *Ann. de Chim. et de Phys.*, 8, p. 275, 1843.

of course, is unsatisfactory, inasmuch as the observations cannot be made at the time when the pressure is applied.

The following table will show the relative decrease in volume for the different gases when subjected to pressure : —

Substance	Pressure	Ratio of Original Volume to Final Volume
Oxygen	83 atmospheres	90 : 1
Ethylene	124 "	356 : 1
Nitric oxide	165 "	251 : 1
Carbon monoxide	165 "	180 : 1
Oxygen	165 "	160 : 1
Fluosilicon	105 "	350 : 1

Aimé thought that the ethylene and fluosilicon were condensed to the liquid state. The evidence, however, is not conclusive. Hydrogen and nitrogen were subjected to a pressure of 220 atmospheres, without showing any indications of liquefaction. The observations seem to show deviations from Boyle's law, but the results obtained for oxygen are somewhat contradictory.

Experiments of Faraday

In 1845 Faraday[1] published a second series of observations. In the introduction, he says : "The

[1] *Phil. Trans.*, 135, p. 155; also, *Alembic Club Reprints*, No. 12, p. 33.

experiments formerly made on the liquefaction of gases, and the results which from time to time have been added to this branch of knowledge, especially by Thilorier, have left a constant desire on my mind to renew the investigation. This, with considerations arising out of the apparent simplicity and unity of the molecular constitution of all bodies when in the gaseous or vaporous state, which may be expected, according to the indications given by the experiments of Cagnaird de la Tour, to pass by some simple law into the liquid state, and also the hope of seeing nitrogen, oxygen, and hydrogen, either as liquid or solid bodies, and the latter probably as metal, have lately induced me to make many experiments on the subject."

The gases, in these experiments, were subjected to the simultaneous influence of high pressure and low temperature. The pressure was obtained by the use of two air-pumps. "The first pump had a piston of an inch in diameter, and the second a piston of only half an inch in diameter, and these were so associated by a connecting pipe, that the first pump forced the gas into and through the valves of the second, and then the second could be employed to throw forward this gas, already compressed to 10, 15, or 20 atmospheres, into its final recipient at a much higher pressure."

The pressure exerted by certain gases when generated in closed, strong glass vessels was also employed as a means of compression.

The low temperatures were produced by means of a bath of Thilorier's mixture of solid carbonic acid and ether. This mixture was allowed to evaporate, at the ordinary atmospheric pressure, in an open earthenware dish which rested in a second larger vessel; the space between the two being filled with dry flannel. Faraday modified the method, however, and obtained a much lower temperature by evaporating the mixture under reduced pressure.

The influence of pressure is shown by the following table : —

Pressure in Inches of Mercury	Temperature of Mixture
28.4	$-$ 70°
19.4	— 80.3
9.4	— 85
7.4	— 88.3
5.4	— 90.6
3.4	— 95
2.4	— 98.3
1.4	— 106.6
1.2	— 110 = — 166° F.

The temperatures were measured by means of an alcohol thermometer, and Faraday remarks that the actual temperatures were probably from 5 to 6° lower than those recorded.

The gases were condensed in tubes, the forms of which are shown in figures 6 and 7. "These tubes were of green bottle glass, being from $\frac{1}{6}$ to $\frac{1}{4}$

FIG. 6.

of an inch in external diameter, and from $\frac{1}{42}$ to $\frac{1}{30}$ of an inch in thickness. They were chiefly of two kinds, about 11 and 9 inches in length, the one, when horizontal, having a curve downward near the end to dip into the cold bath, and the other, being in form like an inverted siphon, could have the bend cooled in the same manner when necessary. Into the straight part of the horizontal tube, and the longer leg of the siphon tube, pressure-gauges were introduced when required. Caps, stop-cocks, and connecting pieces were employed to attach the glass tubes to the pumps, and these, being of brass, were of the usual character of those employed for

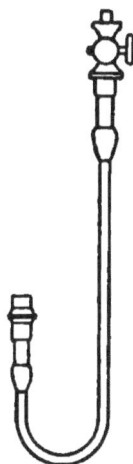

FIG. 7.

operations with gases, except that they were small and carefully made."

These tubes were tested by means of a hydro-

static press.   One of them sustained a pressure of
118 atmospheres without breaking, while another
burst at a pressure of 67 atmospheres.

The pressures were measured by means of a
small glass tube which contained a movable cylinder of mercury, one end of the tube being closed.
A pressure of 10 or 20 atmospheres would reduce
the volume of air in the gauge to $\frac{1}{10}$ or $\frac{1}{20}$ of its
volume at the ordinary atmospheric pressure.
After the tube had once been calibrated, it was
a simple matter to measure the various pressures.

The solid carbonic acid was preserved for experimental purposes by placing it in a glass vessel,
which rested in the middle of three concentric glass
jars, separated from each other by dry jackets of
woollen cloth.   This method proved to be very
effectual.   The results obtained are as follows : —

Olefiant Gas

This gas was condensed to a clear, colorless,
transparent liquid, but could not be solidified at
the lowest temperature obtainable by means of the
carbonic acid mixture.   The author made a large
series of observations on the vapor pressures of
this liquid at different temperatures.   He also suggested condensation and evaporation as a method
of purifying the gas.

Hydriodic Acid

This gas was easily condensed by means of the carbonic acid bath. At a temperature of about −51° it became a clear, transparent, colorless solid, which exhibits fissures or cracks similar to those in ordinary ice.

Hydrobromic Acid

This gas became liquid at a temperature of about −73°. The fluid was colorless and transparent. At and below −87° it is a solid, transparent, crystalline body. The liquid does not freeze until reduced much lower than the above-mentioned temperature, but being frozen by the carbonic acid bath, it remains a solid until the temperature, in rising, attains to −87°.

Fluosilicon

At a pressure of 9 atmospheres and a temperature of −107°, this substance condensed to a liquid. In this condition it was clear, transparent, colorless, and very fluid, like hot ether. It could not be solidified at the lowest temperature employed in these experiments.

Phosphine and Arsine

Phosphine was liquefied, but could not be solidified. Faraday remarks that in this experiment

E

he was able to condense only a portion of the gas, and suggests the presence of another gas which could not be condensed under the conditions of the experiment. This gas, he says, is probably another phosphuretted hydrogen, or hydrogen itself.

In regard to arsine, he says, "this body, liquefied by Dumas and Soubeiran, did not solidify at the lowest temperature to which I could submit it, *i.e.* not at 110° below zero."

Fluoboron

This substance was prepared from fluorspar, fused boracic acid, and strong sulphuric acid, in a tube generator, and conducted into a condensing tube under the generating pressure. The ordinary carbonic acid bath did not condense it, but the application of one cooled under the air-pump produced liquefaction. Fluoboron then appeared as a very limpid, colorless, clear liquid, showing no signs of solidification. At the lowest temperature it appeared as mobile as hot ether.

Muriatic, Sulphurous, and Carbonic Acids

All attempts to solidify muriatic acid were unsuccessful.

Sulphurous acid, at −76°, became a crystalline,

transparent, colorless solid, which sank freely into the liquid.

In regard to carbonic acid, Faraday says, "The solidification of carbonic·acid by Thilorier is one of the most beautiful experimental results of modern times. When the acid is melted and re-solidified by a bath of low temperature, it appears as a clear, transparent, crystalline, colorless body, like ice; so clear, indeed, that at times it was doubtful to the eye whether anything was in the tube."

Euchlorine and Sulphuretted Hydrogen

Euchlorine was easily condensed to a crystalline solid, the color and appearance of which were similar to those of potassium bichromate; it was moderately hard, brittle, and translucent. The solid melted at a temperature of −60°. In attempting to resolidify the substance, a peculiar phenomenon was observed. The author remarks: "Some hours after, wishing to solidify the same portion of euchlorine which was then in a liquid state, I placed the tube in a bath at −80°, but could not succeed either by continuance of the tube in the bath, or shaking the fluid in the tube; but when the liquid euchlorine was touched by a platinum wire it instantly became solid, and exhibited all the properties before described. There are many similar instances among ordinary substances, but the effect in this case makes me hesitate in concluding that all gases which as yet have refused to solidify at temperatures as low as $-110°$, cannot acquire the solid state at such a temperature."

Sulphuretted hydrogen, at a temperature of $-85°$, solidified to a white, crystalline, translucent substance, which was made up of a confused mass of crystals.

Ammonia and Cyanogen

Ammonia was condensed to a white, translucent, crystalline solid, which melted at a temperature of $-85°$.

Cyanogen was also obtained as a transparent crystalline solid, the specific gravity of which was nearly the same as that of the liquid.

Nitrous Oxide

This gas was solidified, at a temperature of $-100°$, to a beautiful, clear, crystalline, colorless body. The pressure of the vapor rising from the solid was less than one atmosphere. Hence "it was concluded that liquid nitrous oxide could not freeze itself by evaporation into the atmosphere, as carbonic acid does; and this was found to be true." Natterer observed the same fact a year earlier. Faraday called attention to the fact that very low temperatures could be obtained by the evaporation of nitrous oxide. He says: "It is probable that this substance may be used in certain cases, instead of carbonic acid, to produce degrees of cold far below those which the latter body can supply. . . . There is no doubt that the substance placed in vacuo would acquire a temperature lower than any as yet known."

Faraday made a large series of observations on the vapor pressures of these condensed gases at different temperatures. The following table contains one result from each of the different substances : —

Substance	Temperature	Pressure
Olefiant gas	0°	43.4 atmospheres
Hydriodic acid	0°	3 97 "
Fluosilicon	0°	30.0 "
Fluoboron	55°	10.0 "
Muriatic acid	0°	26.2 "
Sulphurous acid	0°	1.53 "
Sulphuretted hydrogen	0°	10 0 "
Carbonic acid	0°	38.5 "
Cyanogen	0°	2.37 "
Ammonia	0°	4.44 "
Arsine	0°	8.95 "
Nitrous oxide	0°	32.0 "

The following substances could not be solidified at a temperature of −110° : chlorine, ether, alcohol, carbon disulphide, caoutchoucine, camphine, and oil of turpentine.

Faraday also attempted to liquefy the so-called permanent gases by subjecting them to low temperatures and high pressures, but without success. At the lowest temperature obtainable by means of the carbonic acid mixture, these gases were subjected to the pressures expressed in the following table : —

	Atmospheres
Hydrogen	27
Oxygen	58
Nitrogen	50
Nitric oxide	50
Carbon monoxide	40
Coal gas	32

In no case were there any signs of liquefaction. The author says : " Thus, though as yet I have not condensed oxygen, hydrogen, or nitrogen, the original objects of my pursuit, I have added six substances, usually gaseous, to the list of those that could previously be shown in the liquid state, and have reduced seven to the solid form."

Although Faraday was unsuccessful in his attempts to liquefy the so-called permanent gases, yet he predicted that if the temperature could be sufficiently reduced, these gases would pass into the liquid or solid state. The behavior of gases and vapors at different temperatures, he says, "gives great encouragement to the continuance of those efforts which are directed to the condensation of oxygen, hydrogen, and nitrogen, by the attainment and application of lower temperatures than those yet applied. If to reduce carbonic acid from the pressure of two atmospheres to that of one, we require, then, to abstract only about half the number of degrees that is necessary to produce the same effect with sulphurous acid, it is to be expected that a far less abstraction will suffice to produce the same effect with nitrogen or hydrogen, so that further diminution of temperature, and improved apparatus for pressure, may very well be expected to give us these bodies in the liquid or solid state."

After such an elaborate series of observations, it is not surprising to find that Faraday anticipated the work of Andrews on critical constants. He makes the following statement: "M. Cagnaird de la Tour has shown that at a certain temperature a liquid, under sufficient pressure, becomes clear, transparent vapor, or gas, having the same bulk as the liquid. At this temperature, or one a little higher, it is not likely that any increase of pressure, except perhaps one exceedingly great, would convert the gas into a liquid. Now the temperature of 110° below zero, low as it is, is probably above this point of temperature for hydrogen, and perhaps for nitrogen and oxygen, and then no compression, without the conjoint application of a degree of cold below that we have as yet obtained, can be expected to take from them their gaseous state. Further, as ether assumes this state before the pressure of its vapor has acquired 38 atmospheres, it is more than probable that gases which can resist the pressure of from 27 to 57 atmospheres at a temperature of − 110° could never appear as liquids, or be made to lose their gaseous state at ordinary temperatures. They may probably be brought into the state of very condensed gases, but not liquefied."

EXPERIMENTS OF NATTERER

In 1844 Natterer [1] published an article on the liquefaction and solidification of carbonic acid, nitrous oxide, etc. Historically considered, this work should precede the experiments of Faraday, which have just been outlined. The observations of Natterer, however, extend over a period of a

---

[1] *Jour. Prakt. Chem.*, 31, p. 375; *Pogg. Ann.*, 62, p. 132, 1844.

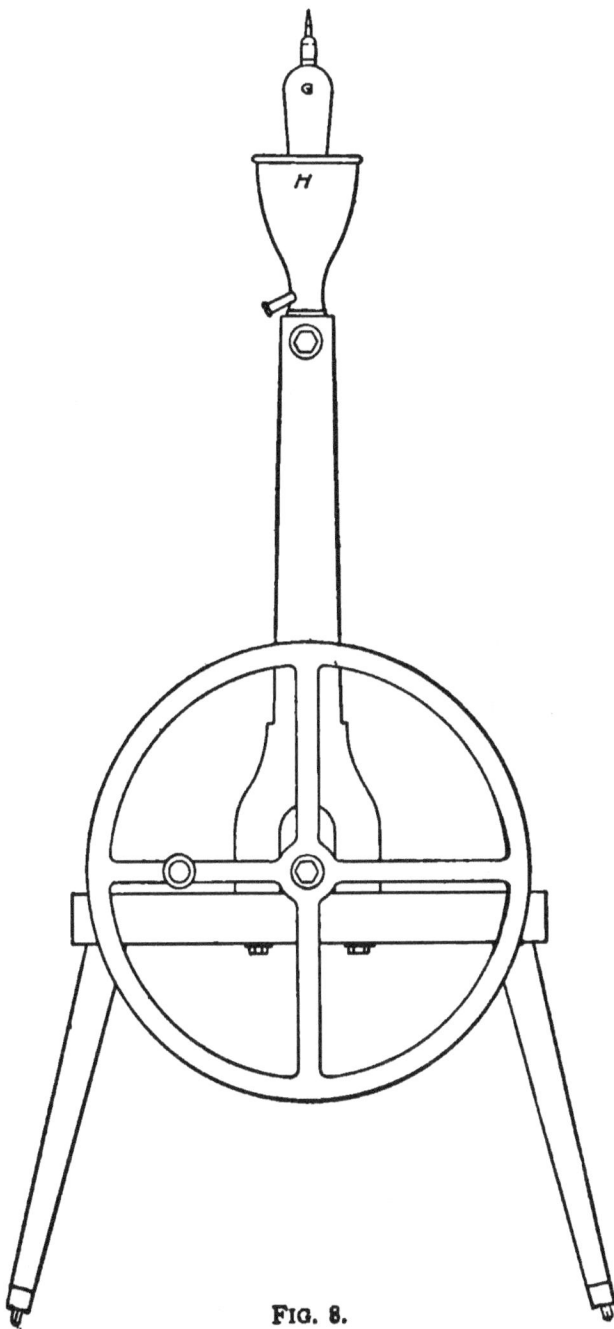

FIG. 8.

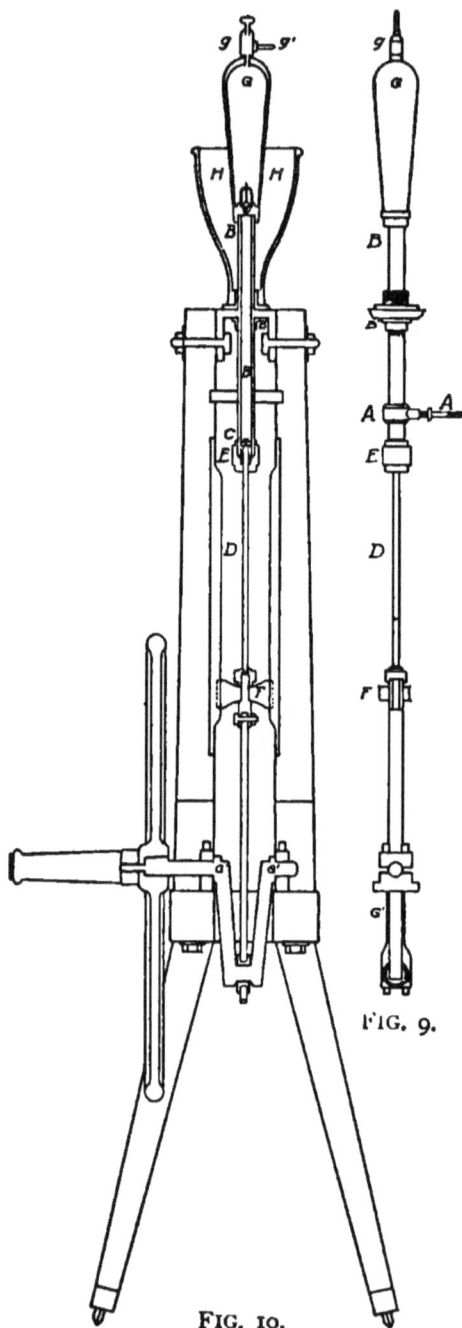

FIG. 10.

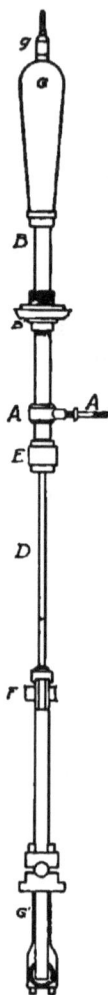

FIG. 9.

number of years, and for that reason are taken up at this point.

The methods employed in the liquefaction of gases were materially advanced by this investigator. He placed in the hands of the experimenters a method for obtaining extremely high pressures. He also introduced a convenient vessel for preserving liquefied gases.

A sketch of the complete apparatus[1] employed by Natterer is shown in figure 8. The details can be

[1] *Jour. Prakt. Chem.*, 35, p. 169.

seen from figures 9 and 10. The tube A (Fig. 9) is provided on one end with a screw. This end of the tube is screwed air-tight into an opening in the cylinder B of the pump, and is fastened in that position by means of the ring A'. The other end of the tube A is provided with notches, so that it can be fastened to a flexible leather or rubber sack. This sack is connected with a receiver, in which the gas to be compressed is kept. The cylinder B of the pump is an iron tube, about 20 inches in length and 0.6 inches in diameter. It consists of two tubes screwed into a larger short section B'. In the interior of the cylinder is a piston C (Fig. 10), provided with a leather cap, and fastened to the piston rod D. On turning the contrivance G', the piston is moved up and down. At the lowest position the piston is below the inlet tube A. The wrought-iron vessel G is provided at the bottom with a valve which opens inward, and at the top with an outlet cock g. The vessel G has a capacity of about 20 litres. A freezing mixture is placed in the chamber H. The receivers were always tested, before use, to a pressure of 150 atmospheres.[1]

---

[1] The apparatus of Natterer has been improved by Bianchi in Paris, and later by Ritchie in Boston.

Condensation of Carbonic Acid

The gas to be liquefied was first dried by passing through a bottle containing calcium chloride. The pump was then started, and the cock g opened. When all the air had been removed from the pump and receiver, the cock was closed and the compression begun. The cooling was accomplished by means of ice water in the vessel H. The operation required from one to one and a half hours. In this way Natterer obtained liquid carbonic acid in considerable quantity. The liquid could be kept in the receiver for several months. He also obtained large quantities of solid carbonic acid and studied its properties.

Nitrous Oxide

Natterer was the first to obtain large quantities of liquid nitrous oxide. He found the boiling point to be $-105°$. He also solidified the gas and filtered the solid out of the liquid. The freezing point was found to be $-115°$.

Experiments on Air, Oxygen, Hydrogen, etc.[1]

Natterer subjected carbon monoxide to a pressure of 150 atmospheres, but observed no indi-

[1] *Wien. Ber.*, 5, p. 351; 6, pp. 557 and 570; 12, p. 199.

cations of liquefaction. This induced him to construct an apparatus, by means of which a pressure of 1000 atmospheres could be obtained. With this apparatus he hoped to condense the so-called permanent gases to the liquid state. This hope, however, was not realized. Only negative results were obtained. He then modified the apparatus, and subjected the gases to pressures as high as 3600 atmospheres (54,000 pounds per square inch), and concluded that with pressure alone these gases could not be liquefied. He next endeavored to combine the high pressure with low temperature, but without success. The low temperature was obtained by means of solid carbonic acid and ether.

The experiments of Natterer, although unsuccessful as far as the liquefaction of the permanent gases is concerned, show that these gases under high pressures deviate considerably from Boyle's law. Further reference to these variations will be made at the end of the chapter.

EXPERIMENTS OF BERTHELOT

In 1850 Berthelot[1] constructed an apparatus, by means of which gases could be liquefied without danger. The pressure was obtained by the

Compt. rend., 30, p. 666.

dilation of a liquid. The apparatus consisted simply of a thick-walled glass tube which was closed at one end. The tube was filled with pure dry mercury free from air, and then drawn out into a capillary at the open end. The large end of the tube was placed in a water bath, and the capillary portion in a current of the gas to be compressed. The water bath was then heated, and a portion of the mercury expelled. On cooling, the mercury contracted, and the portion which had been expelled was replaced by the gas. The capillary portion of the tube was then sealed, and the water bath again heated.

Berthelot succeeded, with tubes of this kind, in condensing chlorine, ammonia, and carbonic acid to the liquid state. He observed the abnormal expansion of liquid carbonic acid, which was pointed out by Thilorier. Berthelot also endeavored to liquefy nitric oxide, carbon monoxide, and oxygen. The results in these cases, however, were negative. The pressure obtained in the experiment on oxygen was about 780 atmospheres. The temperature of the gas in this case was lowered by means of solid carbonic acid. Berthelot came to the conclusion that pressure would not produce liquefaction under all conditions of temperature.

OBSERVATION OF DRION

In 1859 Drion[1] made some observations similar to those made by Andrews a few years later. The experiments, however, were not very elaborate. Working with hydrochloric ether, nitrogen tetroxide, and sulphurous acid, he found that the coefficient of expansion of these substances in the liquid state increases very rapidly with increasing temperature.

He also observed that, when the liquids were gradually heated in closed tubes to the temperature of complete vaporization, the surfaces lost their definition and reflecting power, and became nebulous in appearance. The nebulous zone increased in size both upward and downward, and at the same time became less distinct until the tube appeared entirely empty. Similar appearances were produced in the inverse order by gradually cooling the tube.

When a capillary tube was partially immersed in the liquid, the curvature of the meniscus and the capillary elevation gradually decreased as the temperature rose, and finally, just before complete vaporization, the surface became plane, and the level of the liquid within the tube was the same as that of the liquid without the tube.

[1] *Ann. de Chim. et de Phys.* [3], pp. 5 and 221.

Later, Drion and Loir [1] described a method for liquefying gases in considerable quantity. They lowered the temperature by the method of Bussy; *i.e.* by the evaporation of volatile liquids. These experimenters, however, modified the method by conducting finely divided currents of dry air through the liquid during evaporation. In this way Drion and Loir succeeded in liquefying a number of gases, and studied their properties. The low temperatures were measured by means of an alcohol thermometer.

During the following year these experimenters [2] succeeded in obtaining a perfectly transparent mass of solid carbonic acid. The temperature was lowered by the evaporation of liquid sulphurous acid and ammonia. The transparent mass of carbonic acid, they state, consisted of cubical crystals.

OBSERVATIONS OF MENDELEEFF

In 1861 Mendeleeff called attention to some important relations between the gaseous and liquid states of matter. In the concluding paragraph to an article on " The Expansion of Liquids when Heated above their Boiling Points," [3] he makes the

---

[1] *Phil. Mag.* [4], 20, p. 202, 1860.
[2] *Ibid.* [4], 21, p. 495, 1861.
[3] *Lieb. Ann.*, 119, p. 1, 1861.

following statement: The *absolute boiling point* of a liquid is that temperature at which the cohesion and the heat of vaporization become equal to zero. At this temperature the liquid changes to vapor, regardless of the pressure and volume. The absolute boiling point of alcohol he determined to be 250°, and that of water 580°. The absolute boiling point suggested by Mendeleeff corresponds to the critical temperature introduced by Andrews (see next chapter).

EXCEPTIONS TO BOYLE'S LAW

Before closing the present chapter, reference should be made to some observations on the relation of pressure to the volume of a gas. These experiments were made for most part to test the validity of Boyle's law. Van Marum, in 1799 (see p. 15), called attention to the fact that ammonia gas at high pressures does not *obey* this law. Other isolated experiments have been recorded, but for the present purpose we may begin with the observations of Despretz[1] in 1827. He determined the relative compressibilities of a number of gases. The volumes of the different gases at the beginning of the experiment were equal to each other, but as the pressure was increased, this equality of

[1] *Ann. de Chim. et de Phys.* [2], 34, p. 335.

F

volumes was destroyed. Carbonic acid and ammonia were found to be more compressible than
air. At pressures above 15 atmospheres, he
noticed that the volume of air was slightly less than
that of hydrogen. Similar observations were made
by Pouillet. In 1830 Arago and Dulong[1] experimented with air at pressures ranging from 1 to 27
atmospheres. In these experiments the observed
volume was usually somewhat less than that calculated from Boyle's law.

● In 1854 Natterer studied the effect of exceedingly high pressures on oxygen, nitrogen, air, and
hydrogen. He subjected these gases to pressures
of more than 3000 atmospheres. The apparatus
consisted simply of a compression-pump and a
receiver, which was provided with a manometer.
The quantity of gas which the receiver contained
at the ordinary atmospheric pressure was increased
10, 20, 50, and 100 fold, and so on, and the corresponding pressures measured. The following table[2]
contains some of the results. The numbers in the
first column under each gas represent the volumes
which were forced into the receiver, while those
in the second column represent the corresponding
pressures in atmospheres.

[1] *Ann. de Chim. et de Phys.* [2], 43, p. 74.
[2] *Wien. Ber.*, 12, pp. 204-206, 1854.

HYDROGEN		OXYGEN		NITROGEN	
Volume	Pressure	Volume	Pressure	Volume	Pressure
8	8	7	7	5	5
18	18	17	17	15	15
28	28	27	27	25	25
38	38	37	37	35	35
48	48	47	47	45	45
58	58	57	57	55	55
68	68	67	67	65	65
78	78	77	77	75	75
88	89	87	87	85	85
98	100	97	97	95	96
198	222	117	117	195	206
298	352	147	147	295	336
398	505	177	177	395	542
598	950	187	188	495	882
798	1584	207	210	595	1546
928	2154	507	670	705	2790
1008	2790	657	1354		

A glance at this table will show that the measurements were only approximations. There can
be no doubt from the results, however, that these
gases under high pressures deviate from Boyle's
law. The relative deviations increase with increasing pressure.

Regnault[1] investigated this subject more thoroughly. He introduced the gas into an accurately

[1] *Mem. de l'Acad.*, 21, p 329.

calibrated glass tube, which was kept at a uniform temperature. The gas was then compressed to a certain fraction of the initial volume by means of mercury, and the pressure again measured. If the initial pressure and volume be represented by p_0 and v_0, and the final pressure and volume by p and v, the following table[1] will show that the product of the pressure and volume does not remain constant.

In every case, except hydrogen, the compressi-

AIR		NITROGEN	
p_0	$\dfrac{p_0 v_0}{pv}$	p_0	$\dfrac{p_0 v_0}{pv}$
mm.		mm.	
738.72	1.001414	753.46	1.000988
2112.53	1.002765	4953.92	1.002952
4140.82	1.003253	8628.54	1.004768
9336.41	1.006366	10981.42	1.006456
CARBON DIOXIDE		HYDROGEN	
p_0	$\dfrac{p_0 v_0}{pv}$	p_0	$\dfrac{p_0 v_0}{pv}$
mm.		mm.	
764.03	1.007597	. . .	. . .
3186.13	1.028698	2211.18	0.998584
4879.77	1.045625	5845.18	0.996121
9619.97	1.155865	9176.50	0.992923

[1] Preston, *Theory of Heat*, p. 402.

bility was found to be greater than that calculated from Boyle's law. The compressibility of hydrogen was found to be less than the calculated value. Owing to this peculiar behavior, Regnault distinguished hydrogen as a *gaz plus que parfait.*

During the last thirty years this subject has been very thoroughly investigated by Cailletet, Amagat, and others. The results of these observations will be considered in the next chapter.

CHAPTER III

Conception of Critical Constants

WHILE the investigations on Boyle's law were being carried out, Andrews was considering the problem of the liquefaction of gases. With his observations a new era begins. Maxwell[1] says, "The experiments of Andrews on carbonic acid furnish the most complete view hitherto given of the relation between the gaseous and liquid states of matter, and of the mode in which the properties of a gas may be continuously and imperceptibly changed into those of a liquid." The question as to the relative value of temperature and pressure in the liquefaction of gases had long been discussed. The interesting observations of Caignard de la Tour (p. 18), of Drion (p. 63), and of Mendeleeff (p. 64), all point toward the conclusion that for every gas there is a definite temperature above which it cannot be liquefied. These results, however, remained unheeded until Andrews published

---

[1] *Theory of Heat*, p. 124.

his exhaustive researches in this field. From his observations it is evident that for every gas there is a definite temperature above which it cannot be condensed to the liquid state. This point Andrews called the **Critical Temperature.** The *Caignard de la Tour point,* to which Faraday frequently referred, and the *absolute boiling point* suggested by Mendeleeff, correspond to this same temperature.

Previous to the work on critical constants, Andrews[1] studied the effect of high pressures and low temperatures on the volumes of various gases. He experimented at temperatures varying from $-75°$ to $-110°$. Ordinary air was compressed to $\frac{1}{675}$ of its original volume; in which state the density was but little inferior to that of water. Oxygen was reduced to $\frac{1}{554}$, and hydrogen to $\frac{1}{600}$ of the original volumes. Experiments were also made on carbon monoxide, nitrogen, and nitric oxide. None of the gases showed any indications of liquefaction. ·

The results obtained in the preceding experiments led Andrews to make a careful study of the conditions under which carbonic acid can be liquefied. The first account of this work was communicated by the author to Dr. Miller, who

---

[1] *Chem. News*, 4, p. 158, 1861.

published a short notice of the results in 1863.[1]
The note reads as follows:—

"On partially liquefying carbonic acid by pressure alone, and gradually raising at the same time the temperature to 88° F., the surface of demarkation between the liquid and gas became fainter, lost its curvature, and finally disappeared. The space was then occupied by a homogeneous fluid, which exhibited, when the pressure was suddenly diminished or the temperature slightly lowered, a peculiar appearance of moving or flickering striæ throughout the entire mass. At temperatures above 88° no apparent liquefaction of carbonic acid, or separation into two distinct forms of matter, could be effected, even when a pressure of 300 or 400 atmospheres was applied. Nitrous oxide gave analogous results."

A complete account of the first series of observations was published by the author in 1869.[2] The apparatus employed is shown in figures 11, 12, and 13. The gas to be liquefied was compressed in glass tubes of the form represented in figure 11. The upper portion, from a to b, is a capillary tube. From b to c the diameter is about 2.5 mm., and from c to f about 1.25 mm. Before use the tubes were carefully calibrated by

---

[1] Miller, *Chemical Physics*, 3d. ed., p. 328.
[2] Bakerian Lecture, *Trans. Roy. Soc.*, 1869, Part 2, p. 575.

means of mercury.   The pure, dry, carbonic acid
was conducted through the tube for several hours,
to remove the air.   The capillary end was then
hermetically sealed, and
the other end tempora-
rily closed until the tube
was placed in an upright
position with this end
below the surface of
mercury, as shown in
the third tube (Fig. 11).
Heat was then applied
to the tube until a small
quantity of the gas was
expelled through the
mercury.   On cooling, a
short column of mer-
cury rose in the tube.
The open end, while
under the surface of the
mercury, was then con-
nected with the receiver

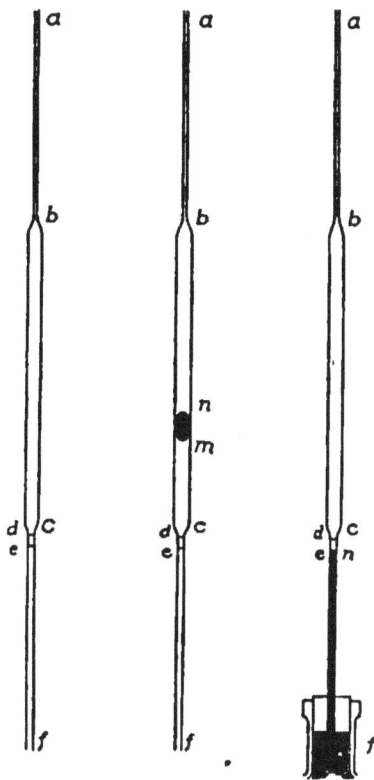

FIG. 11.

of an air-pump, and the tube partially exhausted.
In this way the mercury could be made to rise any
desired height.   The tube being previously calibrated, it was a simple matter then to accurately
calculate the volume of the gas at 0° and 760 mm.
pressure.

The tube containing the gas, with a small quantity of mercury, was placed in a strong, cold-drawn copper tube, provided at each end with a massive brass flange. A longitudinal section of this arrangement is shown in figure 12. Two brass end-

FIG. 12.

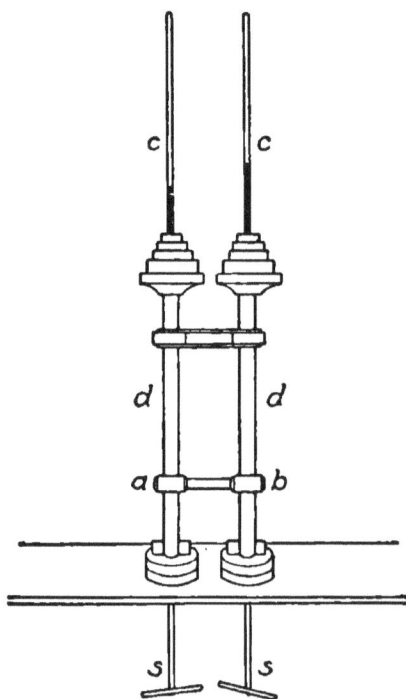

FIG. 13.

pieces were firmly bolted to the flanges, and the connection made air-tight by means of leather washers. The lower end of the tube was provided with a steel screw 180 mm. in length and 4 mm. in diameter. By means of this screw a pressure of 400 atmospheres could be obtained. The portion of the copper tube not occupied by the glass tube was filled with water. When the screw was inserted, the pressure gradually increased; the mercury rose in the glass tube and forced the gas up into the capillary portion which had been calibrated. The complete apparatus is represented in figure 13. One of the tubes contains the carbonic acid, while the other contains air and serves as a comparison tube. A communication is established between the two tubes through a and b, below the mercury, so that the pressures will be equal. The pressures at different intervals were calculated from the relative volumes of the compressed air.

At a temperature of 13°.1, the carbonic acid began to liquefy under a pressure of 48.89 atmospheres. This point could not be observed directly, inasmuch as the first visible traces of liquid represent a column two or three millimetres in length. It was determined indirectly, however; by observing the volume of the gas at a few tenths of a degree above the point of liquefaction, and then calculating the contraction of the gas in cooling to the point where liquefaction begins. The results obtained at the temperatures 13°.1 and 21°.5 are given in the following table, in which δ represents the ratios of the volumes of air before and after compression, ϵ represents the corresponding ratios for carbonic acid, t is the temperature of the air, and t' the temperature of the carbonic acid : —

CARBONIC ACID AT 13°.1				CARBONIC ACID AT 21°.5			
δ	t	ϵ	t'	δ	t	ϵ	t'
$\frac{1}{47.50}$	10°.75	$\frac{1}{76.16}$	13°.18	$\frac{1}{46.70}$	8°.63	$\frac{1}{67.26}$	21°.46
$\frac{1}{48.89}$	10 .86	$\frac{1}{80.90}$	13 .09	$\frac{1}{60.05}$	8 .70	$\frac{1}{114.7}$	21 .46
$\frac{1}{49.00}$	10 .86	$\frac{1}{105.9}$	13 .09	$\frac{1}{60.29}$	8 .70	$\frac{1}{174\ 8}$	21 .46
$\frac{1}{49.28}$	10 .86	$\frac{1}{268.8}$	13 .09	$\frac{1}{60.55}$	8 .70	$\frac{1}{240.5}$	21 .46
$\frac{1}{50.15}$	10 .86	$\frac{1}{462.9}$	13 .09	$\frac{1}{61\ 00}$	8 .70	$\frac{1}{367\ 7}$	21 .46
$\frac{1}{76.61}$	10 .86	$\frac{1}{500.7}$	13 .09	$\frac{1}{62\ 21}$	8 .70	$\frac{1}{440.0}$	21 .46
$\frac{1}{90.43}$	10 .86	$\frac{1}{510.7}$	13 .09	$\frac{1}{62.50}$	8 .70	$\frac{1}{443\ 3}$	21 .46

In both series of results there is an abrupt change of volume at the point where liquefaction begins. The small increase in pressure, which was necessary to continue the liquefaction, was attributed by Andrews to a small quantity of air in the tube containing the carbonic acid. After the condensation to liquid was complete, the pressure increased rapidly. The high coefficient of expansion of liquid carbonic acid, which had been observed by Thilorier, was confirmed by Andrews.

The following table gives the results obtained at the temperatures 31°.1 and 32°.5 : —

CARBONIC ACID AT 31°.1				CARBONIC ACID AT 32°.5			
δ	t	ϵ	t'	δ	t	ϵ	t'
$\frac{1}{54.79}$	$11°.59$	$\frac{1}{80.55}$	$31°.17$	$\frac{1}{57.38}$	$12°.10$	$\frac{1}{85\ 90}$	$32°.50$
$\frac{1}{58.46}$	11.55	$\frac{1}{90.04}$	31.19	$\frac{1}{71.52}$	12.15	$\frac{1}{140.3}$	32.34
$\frac{1}{62.67}$	11.44	$\frac{1}{103\ 1}$	31.19	$\frac{1}{73.60}$	12.30	$\frac{1}{156.0}$	32.45
$\frac{1}{67.60}$	11.63	$\frac{1}{124.4}$	31.15	$\frac{1}{74\ 02}$	12.30	$\frac{1}{159.9}$	32.46
$\frac{1}{73\ 26}$	11.45	$\frac{1}{169.0}$	31.09	$\frac{1}{76.25}$	12.40	$\frac{1}{191.7}$	32.38
$\frac{1}{73.83}$	13.00	$\frac{1}{174.4}$	31.08	$\frac{1}{78\ 52}$	12.50	$\frac{1}{311.8}$	32.48
$\frac{1}{75.40}$	11.62	$\frac{1}{311.1}$	31.06	$\frac{1}{79.77}$	12.35	$\frac{1}{351.3}$	32.54
$\frac{1}{79.92}$	11.16	$\frac{1}{383.0}$	31.10	$\frac{1}{84.90}$	12.35	$\frac{1}{387.8}$	32.75
$\frac{1}{85\ 19}$	11.45	$\frac{1}{405\ 5}$	31.05				

In these experiments there was no evidence that liquefaction had taken place. There was a sudden decrease in volume at a definite pressure in each series, just as in the preceding observations, but the fluid in this case did not separate into two distinct layers.

Observations were also made at temperatures of 35°.5 and 48°.1. In each case the abruptness of the change in volume becomes less marked as the temperature increases. The results may be best compared by reference to the curves (Fig. 14). The ordinates represent the pressures in atmos-

pheres, while the abscissas represent the corre-
ponding volumes.

The curve for 13°.1 shows that at low pressures

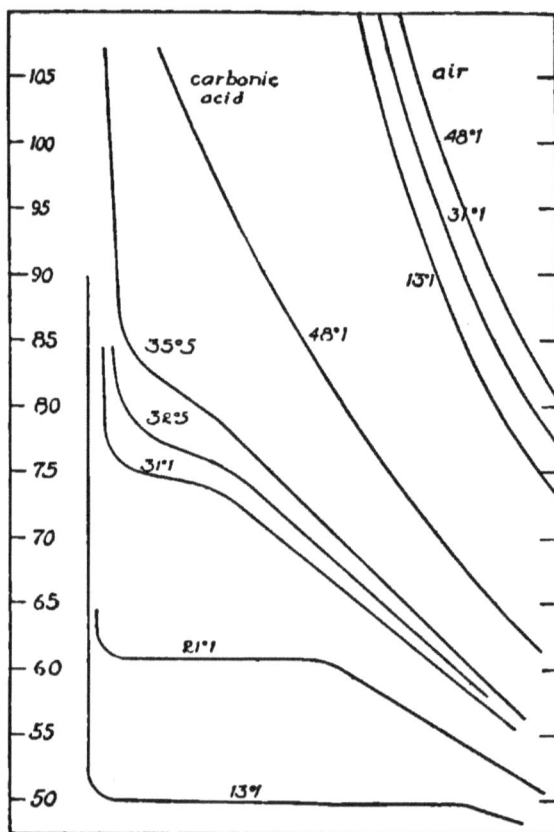

FIG. 14.

the carbonic acid approximately obeys the laws of
gaseous expansion. At a pressure of about 50
atmospheres, the curve becomes almost parallel
to the axis of abscissas; *i. e.* the volume decreases
rapidly while the pressure remains practically con-

stant. This represents the change from the gaseous to the liquid state. When this change is complete, the pressure increases rapidly, and the curve becomes almost parallel to the axis of ordinates. The curve for 21°.5 is very similar to that for 13°.1. The abruptness of the change, however, is not quite so pronounced. The curves for 31°.1 and 32°.5 also show abrupt changes, after which the contraction in volume corresponds approximately to that of liquid carbonic acid. However, liquefaction does not take place at these temperatures. At 48°.1 the curve shows no abrupt changes. Yet at high pressures the gas deviates considerably from Boyle's law, as can be seen by comparison with the air curves in the same figure.

From these results it became evident to Andrews that carbonic acid can be liquefied at or below the temperature of 21°.5, but cannot be liquefied at or above the temperature of 31°.1. He then made a series of careful experiments to determine accurately the highest temperature at which carbonic acid can be liquefied. This was found to be 30°.92 C., or 87°.7 F. Below this temperature two distinct layers of the fluid, separated by a well-defined meniscus, could be observed, while at higher temperatures no such separation could be produced. This point is the critical temperature of carbonic acid. The pressure re-

quired to liquefy carbonic acid at this temperature
is the critical pressure.

The change of state at the critical temperature
is gradual and imperceptible. The author says,
"I have frequently exposed carbonic acid to much
higher pressures, and have made it pass, without
break or interruption, from what is regarded by
every one as the gaseous state, to what is, in like
manner, universally regarded as the liquid state.
Take, for example, a given volume of carbonic
acid gas at 50°, or at a higher temperature, and
expose it to increasing pressure until 150 atmospheres have been reached. When the full pressure has been applied, let the temperature be
allowed to fall till the carbonic acid has reached
the ordinary temperature of the atmosphere. During the whole of this operation no breach of continuity has occurred. It begins with a gas, and
by a series of gradual changes, presenting nowhere any abrupt alteration of volume or sudden
evolution of heat, it ends with a liquid. The
closest observation fails to discover anywhere indications of a change of condition in the carbonic
acid, or evidence at any period of the process of
part of it being in one physical state, and part
in another. That the gas has actually changed
into a liquid would, indeed, never have been sus-
pected, had it not shown itself to be so changed

by entering into ebullition on the removal of the pressure."

Andrews extended his observations to ammonia, ether, carbon disulphide, nitrous oxide, and hydrochloric acid. These substances exhibited properties similar to those of carbonic acid. Their critical temperatures were measured, and found, in some cases, to be above 100°.

Andrews also suggested a distinction between gases and vapors which has met with general approval. This distinction is based upon the critical temperature. Many of the properties of vapors, he says, depend on the gas and liquid being present together. This can happen only at temperatures below the critical point. A vapor, therefore, may be defined as a gas below the critical temperature. A vapor can be condensed to a liquid by pressure alone, while a gas cannot. Carbonic acid, then, is a vapor below 31°, and a gas at higher temperatures. Below 200° ether is a vapor; above that temperature it is a gas.[1]

Determination of Critical Constants

After the introduction of critical constants their measurement became a matter of considerable im-

[1] In 1876, Andrews extended his observations to much higher temperatures, but with similar results. *Trans. Roy. Soc.*, 1876, Part 2, p. 421.

G

portance. Most of the experimenters on the lique-faction of gases have determined critical constants by various methods, and have introduced various definitions for the critical state. The simplest method for determining the critical temperature and critical pressure is that employed by Andrews. The substance is placed in a thick-walled glass tube, which is sealed at one end. The highest temperature at which the liquid meniscus can be formed by pressure is taken as the critical tempera-ture, and the corresponding pressure as the critical pressure.

An ingenious method has been introduced by Cailletet and Colardeau[1] for the determination of the critical constants of substances, such as water, which attack glass at high temperature. The operation is carried out in a steel tube. Numer-ous methods have also been introduced by other experimenters.[2]

Critical Constants of Gaseous Mixtures

Andrews called attention to the fact that the presence of the so-called permanent gases lowered the critical temperature of such gases as carbonic acid. He was unable to liquefy a mixture of three

[1] *Ann. de Chim. et de Phys.* [6], 25, p. 519.
[2] See Heilborn for compilation of literature on critical constants. *Zeit. phys. Chem.*, 7, p. 601.

parts of carbonic acid with four parts of nitrogen at temperatures above $-20°$. The critical tem- perature of carbonic acid containing one-tenth of its volume of air or nitrogen was found to be con- siderably lower than that of the pure gas.

Dewar[1] made a large series of observations on the liquefaction of mixtures of carbonic acid with various substances. He thought that the carbonic acid probably united with the substance to form a definite compound which decomposed when the pressure was released.

Pawlewski[2] worked on organic compounds, and concluded that, for the same class of compounds, the critical point of the mixture lay between those of the two constituents; and he thought that the interval was divided proportionally to the quantities of the compounds present. Ansdell,[3] on the other hand, maintained that this law does not hold true.

During recent years a large amount of work has been done in this field. The references to these investigations, however, will be sufficient for the present purpose. See especially Dr. Kuenen, *Zeit. phys. Chem.*, 11, p. 33; and *Proc. Phys. Soc.*, Lond., 13, p. 523, 1895, and 15, p. 235, 1897;

[1] *Proc. Roy. Soc.*, 1880.
[2] *Ber. d. d. chem. Ges.*, 1882, p. 460.
[3] *Proc. Roy. Soc.*, 34, p. 113, 1882.

Duhem, *Jour. Phys. Chem.*, 1, p. 273, 1897. Theoretical, Van der Waals, *Zeit. phys. Chem.*, 5,
p. 133.

SECTION II

*Continuity of the Gaseous and Liquid States of
Matter*

Since the introduction of critical constants by
Andrews much discussion has arisen concerning
the condition of matter at or near the critical
point. Andrews was the first to call attention to
this problem. "What is the condition of carbonic
acid," he asks, "when it passes at temperatures
above 31°, from the gaseous state down to the
volume of the liquid, without giving evidence at
any part of the process of liquefaction having
occurred? Does it continue in the gaseous state,
or does it liquefy, or have we to deal with a new
condition of matter? If the experiment were made
at 100°, or at a higher temperature, the probable
answer would be that the gas preserves its gaseous
condition during the compression. On the other
hand, when the experiment is made at temperatures a little above 31°, the great fall which occurs
at one period of the process would lead to the con-
jecture that liquefaction had actually taken place,
although optical tests carefully applied failed at

any time to discover the presence of liquid in contact with a gas. . . . Under many of the conditions which I have described it would be vain to
attempt to assign carbonic acid to the liquid rather
than the gaseous state. Carbonic acid, at the temperature of $35°.5$, and under a pressure of 108
atmospheres, is reduced to $\frac{1}{450}$ of the volume
occupied under a pressure of one atmosphere; but
if any one ask whether it is now in the gaseous or
liquid state, the question does not, I believe, admit
of a positive reply. The substance in this condition stands nearly midway between the gas and the
liquid."

Numerous observations have been made by later
experimenters to show that the liquid state persists beyond the critical point, while other investigators have endeavored to show that this point is
the limit of the liquid state.

Avenarius[1] made a series of observations on
"The Causes which Determine the Critical Point,"
and concluded that the volume of the saturated
vapor at the critical point is not equal to that of
the liquid at the same temperature. He experimented with ether.

Ansdell,[2] on the other hand, experimenting with

[1] *Mem. Acad. Sci.*, St. Petersburg, 1876–1877; Ref., Ansdell,
Proc. Roy. Soc., 30, p. 120.

[2] *Proc. Roy. Soc.*, 30, p. 117, 1880.

hydrochloric acid, says, "The volumes of the saturated vapor and liquid gradually approach each
other as the temperature nears the critical point,
and would undoubtedly become identical if the
experiments could be carried on up to the critical

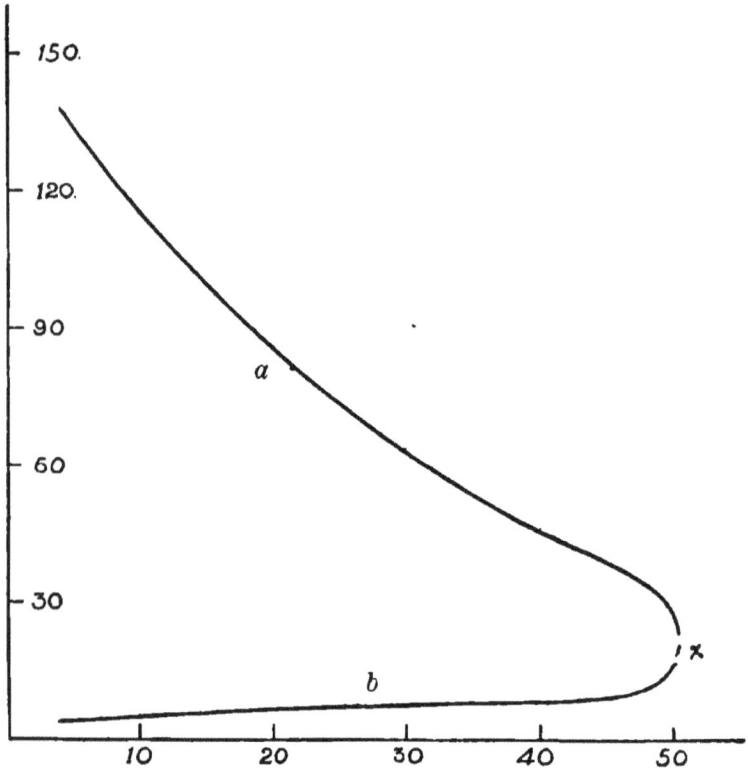

FIG. 15.

point." The results of his experiments are shown
by the curves in figure 15. The volumes are represented by ordinates, and the temperatures by
abscissas. The curve *a* represents vapor volumes,

and the curve *b* liquid volumes. At a temperature slightly above 50°, the curves show considerable change in direction. Both approach the critical point *x*, but the experiments could not be successfully carried to this temperature.[1]

Ramsay[2] says, "The critical point is that point at which the liquid, owing to expansion, and the gas, owing to compression, acquire the same specific gravity, and consequently mix with each other."

During this same year, Clark[3] made a series of experiments with sulphurous acid in the two branches of a U-tube. He found that when the temperature was near the critical point, the disappearance of the meniscus in one branch did not affect the level of the liquid in the other. From this he concluded that the density of the vapor at the critical point is equal to that of the liquid.

This theory seems to have been generally accepted. Thiesen[4] says, "A substance is to be considered a liquid, or gas, according as the density is greater or less than the critical density."

---

[1] Amagat has investigated this subject more thoroughly, and says the density of the vapor rapidly approaches that of the liquid at the critical temperature. *Compt. rend.*, 114, p. 1093. See also Cailletet and Mathias, *Jour. de Phys.* [2], 5, p. 549, 1886.

[2] *Proc. Roy. Soc.*, 30, p. 323, 1880.

[3] *Proc. Phys. Soc.*, Lond., 4, p. 41, 1880.

[4] *Zeit. compr. und flüss. Gase*, 1, p. 87, 1897.

Quite recently, however, von P. de Heen-Lüttich
has published an article entitled, " Über die ange-
blichen Anomalieen in der Nähe des kritischen
Punktes," [1] in which he states that there are two
critical densities, — the critical density of the liquid
and the critical density of the vapor. The latter,
he says, is equal to one-half of the former. A
large number of observations were made, and the
results obtained seem to contradict some of the
work of previous experimenters.[2]

The condition of a substance at or near the
critical point, however, has been more thoroughly
investigated from a different standpoint. Hannay
and Hogarth [3] examined the solvent properties of
some fluids for non-volatile solids during the passage of the solvent through the critical point.
A precipitation of the solid, on passing through
this point, they thought, would be conclusive evidence of a change from liquid to gas. If, on the
other hand, the solid remained in solution, it would
be an evidence of the perfect continuity of the
gaseous and liquid states.

In the first experiment potassium iodide was
dissolved in alcohol, and the solution heated in a
closed glass tube. There was no indication of the

[1] *Zeit. compr. und flüss. Gase*, 2, pp. 97, 113, and 134, 1898.
[2] See also M. Gouy, *Compt. rend.* 115, p. 720, 1892.
[3] *Proc. Roy. Soc.*, 30, p. 178, 1880.

separation of a solid, even at a temperature of
350°, which is nearly 100° above the critical point
of alcohol.

The next experiment consisted in dissolving the
crystals of potassium iodide in alcohol vapors.
The temperature of the alcohol was never lower
than 65° above the critical point. When the pressure was strongly increased, the potassium iodide
dissolved. By releasing the pressure rapidly, fine
crystals of the salt separated out as a film on the
glass, and in some cases as a cloud of fine particles,
which floated about in the menstruum.

Other solutions were then examined. Calcium
chloride in alcohol and sulphur in carbon disulphide, showed no indications of precipitation at
temperatures above the critical points of the
solvents. The blue solution of cobaltous chloride
in alcohol preserved its color at temperatures far
above the critical point of alcohol. The absorption
spectrum showed no change on passing through
the critical point. The authors accepted these
results as evidence of the solubility of solids in
gases, and of the perfect continuity of the gaseous
and liquid states. Ramsay considers the last of
these experiments as an evidence that the liquid
state persists beyond the critical point. He says:
" Messrs. Hannay and Hogarth found that the
absorption spectrum of colored salts remains un-

altered even when the liquid in which they are dissolved loses its meniscus. Surely no clearer proof is needed to show that the solids are not present as gases, but are simply solutions in a liquid medium." [1]

Ramsay [2] has determined the condition of a fluorescent solution at temperatures above the critical point of the solvent. Eosine in alcoholic solution failed to show the phenomenon of fluorescence after the meniscus had disappeared. The author showed that benzene, at temperatures slightly above the critical point, expands and contracts, with regard to pressure, approximately the same as a liquid. The gaseous state, he thinks, depends entirely upon the mean velocity of the molecules and not upon the mean free path. He considered a gas as a substance consisting of simple molecules, and a liquid as a substance consisting of aggregates of gaseous molecules. Above the critical point, he says, a substance may be wholly liquid or gas, depending upon the pressure.

Three years later Jamin [3] announced a similar theory. He saw no reason why the liquid state should cease at the critical point. He says: " I believe that gases are liquefiable at any temperature when the pressure is sufficient, but that

[1] *Proc. Roy. Soc.*, 30, p. 327.   [2] *Ibid.*, 31, p. 194.
[3] *Phil. Mag.* [5], 16, p 71, 1883.

an unperceived circumstance has prevented the liquefaction from being seen." He looked upon the disappearance of the meniscus only as an evidence that the density of the liquid was equal to that of the vapor, — a condition which could be brought about by means of pressure alone. In support of this theory the following experiment of Cailletet[1] is quoted: A mixture of one part of air and five parts of carbonic acid was compressed until a portion of the latter gas was liquefied. The pressure was then increased to 200 atmospheres, when the meniscus disappeared. Jamin explains this phenomenon on the assumption that the density of the gas at that pressure was equal to that of the liquid.

Cailletet and Colardeau[2] have also endeavored to prove that the liquid state persists at temperatures above the critical point. They experimented with a solution of iodine in liquid carbonic acid. The colored solution was heated, in a closed glass tube, to the critical point of the solvent. After the meniscus disappeared the portion of the tube which had contained the liquid remained colored, while the upper portion was colorless. The absorption spectrum of the substance in this condition showed that the iodine existed in solution, and

[1] *Compt. rend.*, 90, p. 210, 1880.
[2] *Ann. de Chim. et de Phys.* [6], 18, p. 269, 1889.

not as a vapor. The authors concluded then that the liquid state still persisted in the lower portion of the tube. Villard,[1] however, has shown that under certain conditions iodine is soluble in gaseous carbonic acid, and that this does not serve as a test for the liquid state.

The theory that the critical point is the limit of the liquid state has been ably defended by Hannay. Reference has already been made to his work on the solubility of solids in gases. Later he investigated the phenomenon of capillarity[2] for carbonic acid, ammonia, sulphurous acid, nitrous oxide, carbon disulphide, chlorine, and various alcohols, and found that capillarity disappears at or near the critical point.[3] He looked upon the solvent power of a gas as depending upon two conditions, the *molecular closeness* and the *vis viva*. At constant density, he says, the solvent power increases with the kinetic energy of the molecules. This statement, however, requires some modification. Altschul[4] has carefully studied the influence of temperature on the solvent power of a gas. He filled a strong glass tube with a solution of 1 gram

---

[1] *Ann. de Chim. et de Phys.*, 10, p. 387, 1897.

[2] *Proc. Roy. Soc.*, 30, p. 478, 1880.

[3] Similar observations were also made by Clark. *Proc. Phys. Soc.*, Lond., 4, p. 41.

[4] *Zeit. compr. und flüss. Gase*, 1, p. 207, 1897.

of potassium iodide in 150 grams of alcohol, and carefully heated the substance to the critical point of the solvent. At this temperature the substance remained in solution, but when gradually heated to a temperature of about 356° a portion of the tube became filled with very fine glistening crystals, which could be seen with the naked eye. These crystals did not settle to the bottom, but floated about throughout the whole space of the tube. From this observation, it is evident that the statement of Hannay that, "retaining the volume the same, the higher the temperature the greater the solvent power of a gas," is true only within certain limits.

Hannay also made use of an entirely different method for studying the condition of a fluid at the critical point. The apparatus employed is shown in figure 16. The liquid was enclosed in the upper part of the tube *A* by means of the mercur *B*. The lower part of the appa-

FIG. 16.

ratus *C* was filled with nitrogen. By compressing the mercury, and gently tapping the tube, a small bubble of the gas could be made to pass up through the column of mercury into the liquid.

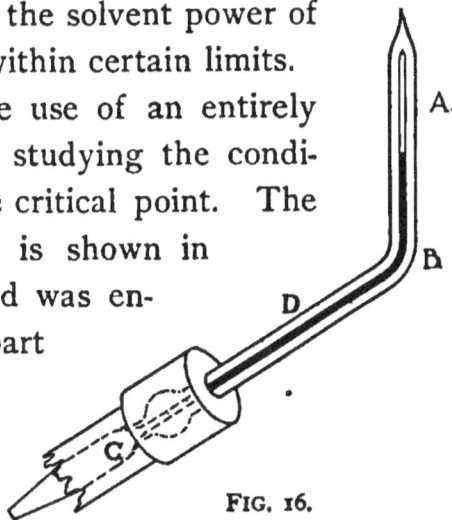

At temperatures below the critical point, the bubble of nitrogen showed a distinct meniscus, and passed up through the column of liquid. At temperatures above the critical point, the bubble diffused immediately throughout the whole space. Hannay looked upon this experiment as decisive evidence that the liquid state ceases at the critical temperature. He says: "There can be no liquid above the critical point, as this is the termination of all properties which distinguish a liquid from a gas."

We come now to the question originally asked by Andrews. "What is the condition of matter at temperatures slightly above the critical point?" Does it belong to the liquid, or the gaseous state? The answer to this question depends entirely upon the definitions of the terms liquid and gas. Hannay[1] suggests four states of matter, — the gas, vapor, liquid, and solid. The vapor, he says, is a distinct state of matter. The fact, however, that some of the properties of vapors differ from those of gases under ordinary conditions, is not necessarily an evidence of a different state of matter, in the sense in which that term is usually employed. Crookes[2] has shown that the properties of gases under extremely low pressures differ

[1] *Proc. Roy. Soc.*, 31, p. 520.
[2] *Ibid.*, 30, p. 469, 1880.

essentially from those of gases under the ordinary pressure, and he has suggested the term *radiant matter* for such conditions. Three states of matter, then, apart from the liquid and solid, have already been suggested, — namely, the vapor, the gas, and radiant matter. It is evident, from the above suggestions, that the state which shall be assigned to matter under definite conditions depends entirely upon the definitions of the various terms. If we limit the states of matter to the solid, liquid, and gas, which is probably advisable, and accept the ordinary definitions, — namely, that a gas has neither form nor volume, but tends to expand indefinitely, and that a liquid has a definite volume, but assumes the form of the vessel in which it is contained, then it seems that a fluid above the critical point belongs to the gaseous state. Matter in that condition tends to expand indefinitely without showing any signs of ebullition. If a small bubble of an indifferent gas is allowed to enter the fluid, it shows no meniscus, but diffuses imperceptibly throughout the whole space.

The gaseous and liquid states, however, gradually approach each other at the critical point. The change from a gas to a liquid at this point seems to be a continuous process, and shows no abrupt evolution of heat. The intimate relation which exists between the two states has been admirably expressed by Andrews. He says: "The ordinary gaseous and ordinary liquid states of matter may be made to pass into one another by a series of gradations so gentle that the passage shall nowhere present any interruption or breach of continuity. From carbonic acid as a perfect gas to carbonic acid as a perfect liquid, the transition we have seen may be accomplished by a continuous process, and the gas and liquid are only distant stages of a long series of continuous physical changes."

SECTION III

Relation between the Gaseous and Liquid States, as expressed by the Equation of Van der Waals

According to Boyle's law (p. 7), the product of the pressure and volume of a gas is a constant quantity. This relation is expressed by the equation,

$$pv = c.$$

According to the law of Charles-Gay Lussac (p. 10), the volume v_1 of a gas at any temperature t, and under the constant p_0, is given by the equation,

$$v_1 = v_0 (1 + at),$$

where v_0 is the volume of the gas at $0°$, and a is the coefficient of expansion $(= \frac{1}{273})$.

If the pressure in the last equation should change from p_0 to any pressure p, the resulting volume v, according to Boyle's law, is,

$$v = v_1 \frac{p_0}{p} = v_0 (1 + at) \frac{p_0}{p};$$

hence, $pv = p_0 v_0 (1 + at)$.

If T represents the absolute temperature, then $t = T - 273$. The coefficient of expansion a being equal to $\frac{1}{273}$, we have,

$$pv = p_0 v_0 \left[1 + \tfrac{1}{273} (T - 273) \right],$$

or, $pv = \dfrac{p_0 v_0}{273} T$.

The expression $p_0 v_0$ represents the product of the pressure and volume at $0°$ and 760 mm. pressure; hence, $\dfrac{p_0 v_0}{273}$ is a constant. Calling this term R, we have $pv = RT$.   I

This is the equation usually employed to express the relation of the pressure and volume to the temperature of a gas. It represents a combination of the law of Boyle with that of Charles-Gay Lussac.

Reference has already been made to the fact that gases under high pressures deviate from Boyle's law (p. 65). The most elaborate series of

H

experiments in this direction is that of Amagat.[1]
He took into account both pressure and temperature, and hence his results apply directly to
equation I.

The results obtained by Amagat are represented
by the curves in figures 17, 18, 19, and 20. Fig-

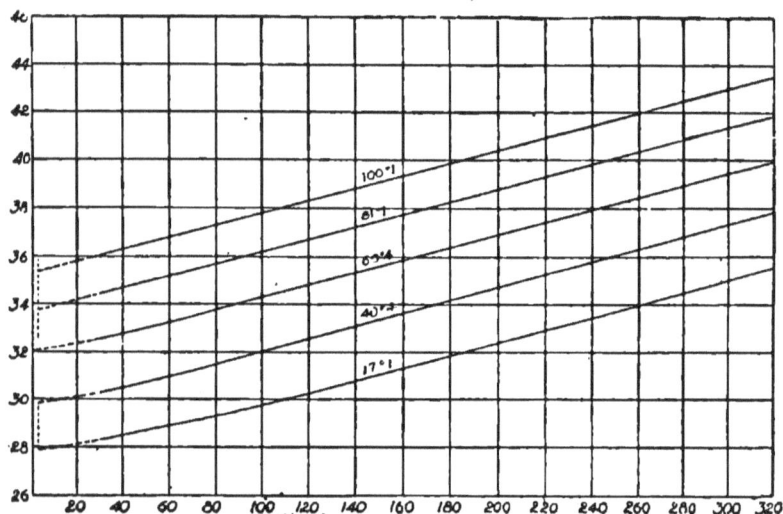

FIG. 17.

ure 17 shows the compressibility of hydrogen.
The abscissas represent the pressures in atmospheres, while the ordinates represent the corresponding values of the product pv. If the equation
$pv = RT$ should remain true under all conditions,
the curves would become straight lines parallel to

[1] *Ann. de Chim. et de Phys.* [4], 29; [5], 23, p. 358; *Compt.
rend.*, 73, p. 183; 75, p. 479.

the axis of abscissas. The curves for hydrogen
are approximately straight lines under the conditions of the experiments, but the value of pv increases, after a certain point has been reached,
with increasing pressure.

The curves for nitrogen are shown in figure 18.
In this case the product of pv decreases slightly

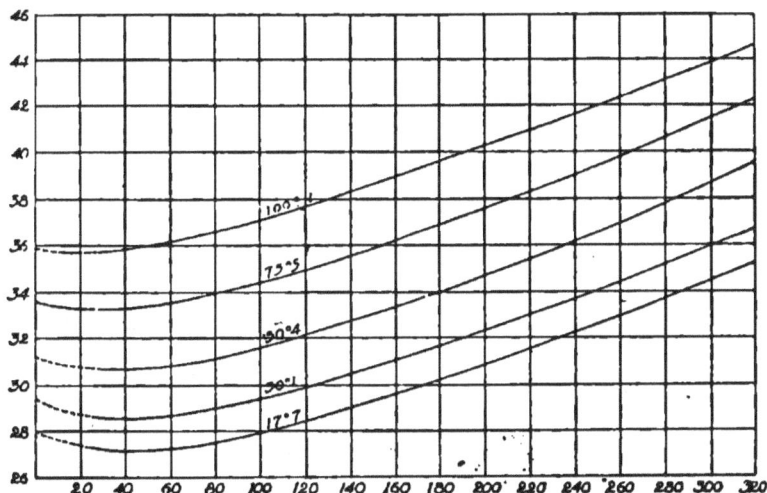

FIG. 18.

at first, and then increases with increasing pressure. As the temperature is lowered the curves
for both hydrogen and nitrogen show greater
divergences from a straight line.

Figure 19 represents the curves for ethylene.
In this case the divergence from a straight line
becomes more pronounced. The value of pv decreases rapidly at first, and then increases. The

change in the direction of the curve becomes more
abrupt as the temperature approaches the critical
point.   The form of the curves for extremely high

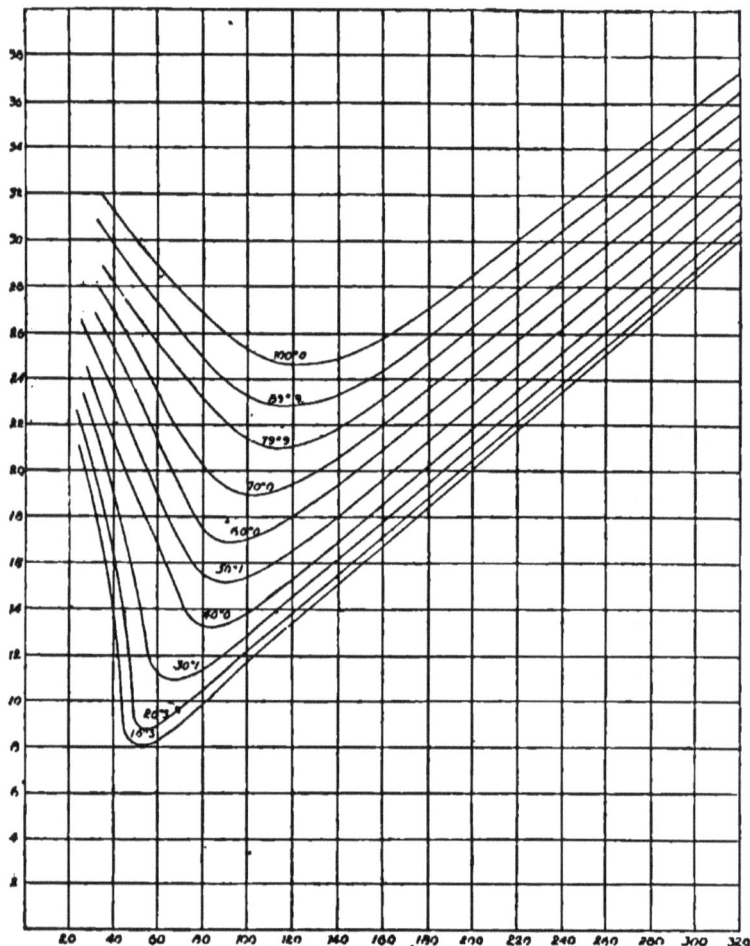

FIG. 19.

temperatures would be similar to those of nitrogen
and hydrogen.

The curves for carbonic acid (Fig. 20) are very

similar to those for ethylene. Near the critical
point the change in the direction of the curve
becomes very abrupt. The compressibility at 35°,

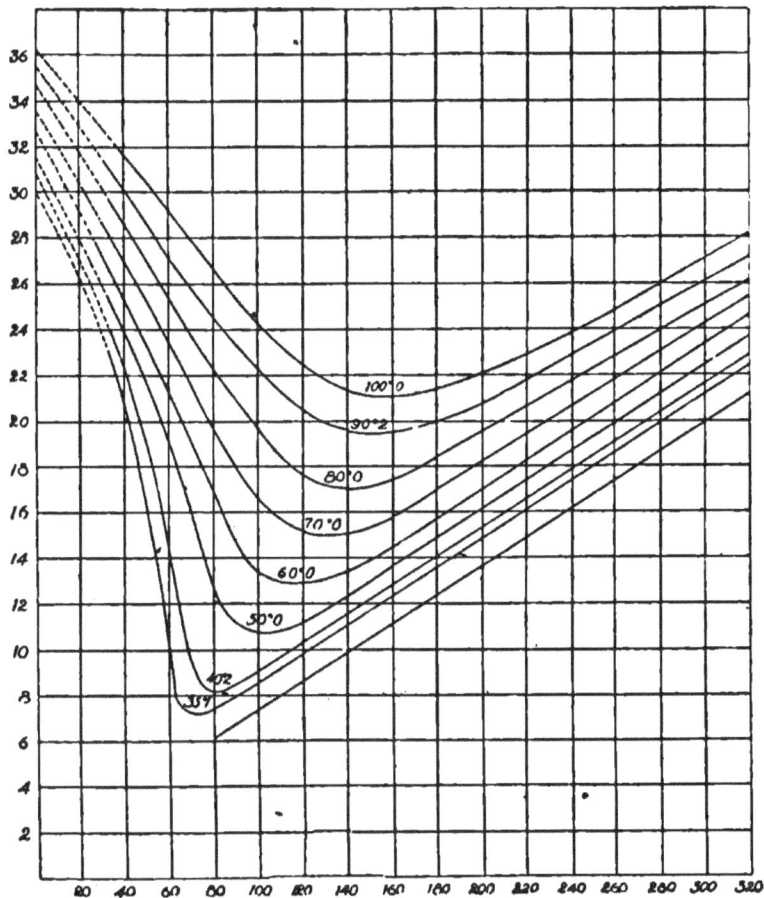

FIG. 20.

and above 80 atmospheres pressure, is approximately the same as for liquid carbonic acid.

In 1888 Amagat[1] extended his observations on

[1] *Compt. rend.*, 107, p. 523.

air, nitrogen, oxygen, and hydrogen to pressures as high as 3000 atmospheres, and found that the compressibility of these gases under extremely high pressures is approximately the same as for liquids. The density of oxygen at a pressure of 3000 atmospheres was found to be greater than that of water.

Reference might be made to other experiments of Amagat in this field, as well as to those of other investigators. The preceding examples, however, are sufficient to show that the equation $pv = RT$, while approximately true under certain conditions of temperature and pressure, cannot be applied to gases under all conditions.

Numerous attempts have been made to deduce a formula which will express the mutual relation of the pressure, volume, and temperature of gases under all conditions. The most complete equations are those of Van der Waals and Clausius. Both of these equations are based upon the kinetic theory of gases, according to which the molecules of all gases are in a state of motion. The pressure of a gas, according to this theory, is due to molecular impacts against the walls of the containing vessel.

The kinetic theory of gases involves the assumption that gaseous molecules are perfectly elastic. Granting this assumption, and supposing each molecule, between any two successive collisions against other molecules or against the walls of the vessel, to move in a straight line, then the pressure of a gas, in terms of molecular velocity, etc., can be calculated. Some of the calculations in this direction are very elaborate, and involve higher mathematics. The more elementary calculation, which can be found in numerous text-books, is sufficient for the present purpose.

Suppose the gas to be contained in a cube, the side of which is l units in length. Let n represent the number of molecules, each of which is moving with the velocity u, and m the mass of each molecule. The velocity u of a molecule moving in any direction can be resolved into the three components, x, y, and z, at right angles respectively to three sides of the cube, provided the relation $x^2 + y^2 + z^2 = u^2$ is preserved.

Suppose we consider a single molecule moving with the velocity x perpendicular to one side of the cube. The molecule, being perfectly elastic, rebounds from the wall of the vessel with a velocity equal to that with which it approaches. Consequently the impulse given to the wall by each collision will be equal to $2\,mx$, — $i.e.$ twice the momentum of the molecule. The collision against the two parallel faces of the cube will take place $\dfrac{x}{l}$

times per unit of time. Hence, the total action of

a molecule per unit of time is $\dfrac{2\,mx^2}{l}$. The same

is true for the other two components, and we have,
for the total action of a molecule against all sides
of the vessel per unit of time, the value,

$$\frac{2\,m}{l}(x^2 + y^2 + z^2) = \frac{2\,mu^2}{l}.$$

The total impulse, then, of all the molecules per

unit of time is equal to $\dfrac{2\,mnu^2}{l}$. This value rep-

resents the total pressure on all sides of the vessel.
Dividing this expression by the area of the cube
($6l^2$), we obtain the pressure p per unit of area;

i.e. $p = \dfrac{2\,mnu^2}{6\,l^3}$. The volume of the vessel, of

course, is equal to l^3. Representing this by v,
and simplifying, we have

$$pv = \tfrac{1}{3}\,mnu^2,$$

or

$$pv = \tfrac{2}{3}\,mn\frac{u^2}{2}.$$

According to this expression, the product of the
pressure and volume of a gas is equal to two-thirds
of the kinetic energy of the molecules. Inasmuch
as the temperature of a gas varies directly with

the kinetic energy of the molecules, this equation is similar in every respect to the equation,

$$pv = RT.$$

The above calculations are based upon the assumption that the molecules are free to move throughout the entire space of the containing vessel, and that the mutual attraction between the molecules is inappreciable. It is evident, however, that the molecules are free to move only in the space which is not occupied by the molecules themselves. A correction must be made, therefore, for the actual volume of the molecules. At low pressures this correction is very small, but at high pressures, where the molecules are crowded together more closely, it becomes appreciable. At high pressures, also, the mutual attraction between the molecules must be taken into account.

In the equation developed by Van der Waals[1] the influence due to the actual volumes of the molecules, and to the mutual attraction between the molecules, is taken into consideration. The complete equation is

$$\left(p + \frac{a}{v^2}\right)(v - b) = RT, \qquad \text{II}$$

in which p, v, R, and T represent the same values

[1] *Die Continuität des gasformigen und flüssigen Zustandes.*

as in equation I (p. 97), a is a function of the mutual attraction between the molecules, and b is proportional to the actual space occupied by the molecules.

The development of this equation represents an elaborate mathematical calculation. For the present purpose, however, we will consider only the application. Taking the atmospheric pressure as unit pressure, and the volume of one gram of gas at 0° and a pressure of one atmosphere as unit volume, Van der Waals calculated the values of a and b, from the compressibility of carbonic acid gas, and found them to be 0.00874 and 0.0023 respectively. Substituting these values, we have

$$\left(p + \frac{0.00874}{v^2}\right)(v - 0.0023) = RT.$$

Placing p and v each equal to unity, the corresponding temperature is 0°, or $T = 273$; and hence,

$$R = \frac{1.00646}{273}.$$

If a and b remain constant under all conditions, we have as a general equation for carbonic acid

$$\left(p + \frac{0.00874}{v^2}\right)(v - 0.0023) = \frac{1.00646\,T}{273}.$$

The values of p and v obtained from this equation for different values of T are represented by

the curves [1] in figure 21.   The volumes are represented by abscissas, and the corresponding pressures by ordinates.   The curve AD corresponds to

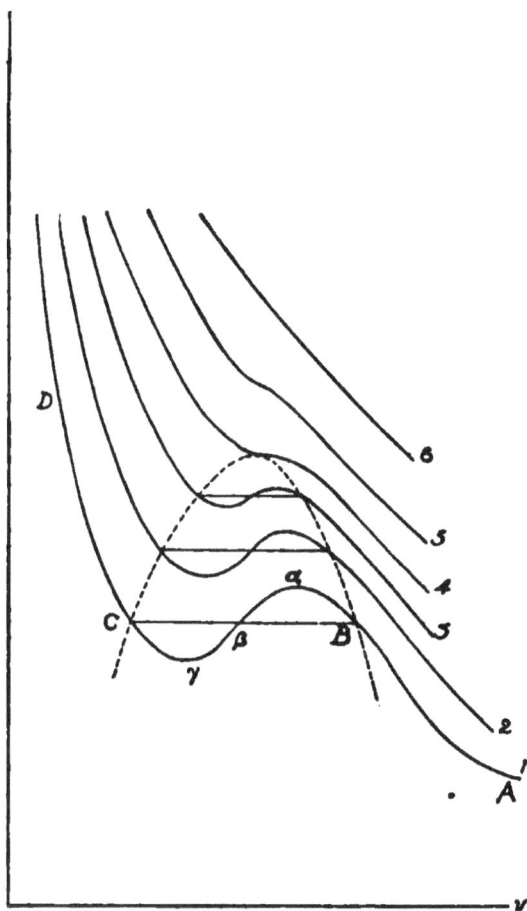

FIG. 21.

a temperature of 13°.1.   The curve, instead of
passing from B to C in a straight line, as was

[1] Ostwald, *Outlines of Gen. Chem.*, p. 86.

observed by Andrews (p. 78), is continuous, and
passes through the points *a* and γ. In 1871
J. Thomson[1] suggested that the curves given by
Andrews could probably be obtained as continuous
curves. The curves represented in figure 21 are
very similar to those suggested by Thomson. In
each of the curves 1, 2, and 3 it will be noticed that
there are three volumes corresponding to one pressure. In curve 1 the three volumes corresponding
to a single pressure occur only between the limits
of pressure represented by γ and *a*. As the temperature increases, the three points on the curves
representing these three different volumes gradually approach each other, and finally meet at the
point *k*.

The point *k*, where the three volumes become
equal to each other, was called by Van der Waals
the critical point. For carbonic acid this point
corresponds to a temperature of 32°.5. Above
this temperature there is but one volume corresponding to a definite pressure. This temperature is about one and a half degrees higher than
the critical temperature obtained experimentally
by Andrews. The pressure corresponding to the
point *k* represents the critical pressure. In the
case of carbonic acid the calculated value is 61

---

[1] *Proc. Roy. Soc.*, 1871.

atmospheres, while the experiments of Andrews show a value of about 70 atmospheres. The volume corresponding to the point k represents the critical volume.

The same results may be obtained by solving the equation algebraically. If the equation be arranged according to the descending powers of v, we have

$$v^3 - \left(b + \frac{RT}{p} \right) v^2 + \frac{a}{p} v - \frac{ab}{p} = 0.^1$$

This equation is of the third degree with respect to v, and, of course, has three roots. A solution of the equation will show that the three roots are real, or that one is real and the other two imaginary. That is, for any pressure there are either three values or one value of v.

This fact is also shown by the graphic representation in figure 21. These curves show that, within a certain interval of pressure for each temperature below 32°.5, there are three corresponding values

---

[1] The equation as given by Van der Waals is

$$v^3 - \left\{ b \frac{(1+a)(1-b)(1+at)}{p} \right\} v^2 + \frac{a}{p} v - \frac{ab}{p} = 0.$$

In this formula, we have, instead of R, the expression $(1+a)(1-b)$. This value is obtained by placing p and v each equal to unity in the original equation of Van der Waals. Substituting R for the expression $(1+a)(1-b)$, and introducing the absolute temperature T, we obtain the equation given above.

for the volume; while above 32°.5, each pressure corresponds to a definite volume. At the critical point k, the three volumes are equal to each other.

If we assume, then, that at the critical point, the three values of v, or the three roots of the equation

$$v^3 - \left(b + \frac{RT}{p}\right)v^2 + \frac{a}{p}v - \frac{ab}{p} = 0$$

are each equal to ϕ, the value of ϕ must be such that

$$3\phi = \frac{b + RT}{p}, \quad 3\phi^2 = \frac{a}{p}, \quad \text{and} \quad \phi^3 = \frac{ab}{p}.$$

Simplifying these expressions, and representing the critical temperature by θ, and the critical pressure by π, we obtain,

$$\phi = 3b = \text{critical volume.}$$

$$\pi = \frac{a}{27\,b^2} = \text{critical pressure.}$$

$$\theta = \frac{8}{27}\frac{a}{Rb} = \text{critical temperature.}$$

Having once determined the values of a and b for a gas, the critical constants can be calculated from the above formulas. The critical constants of carbonic acid calculated in this manner, and the corresponding values obtained experimentally by Andrews, are as follows: —

	Calculated	Experimental
Critical temperature . . .	$32°.5$	$30°.92$
Critical pressure	61 atm.	70 atm.
Critical volume	0 0069	0.0066

The equation of Van der Waals applies to a
homogeneous gas or a homogeneous liquid. It
does not, however, apply to a liquid and its vapor
in mutual contact. During the last twenty years
this equation has been the subject of considerable
discussion. The very elaborate series of experiments by Young [1] show that the generalizations of
Van der Waals can be regarded only as approximations.

In 1880 Clausius [2] called attention to some deviations from the equation of Van der Waals. He
regarded the assumption of Van der Waals, that
the mutual attraction of the molecules is independent of the temperature, and is a function only of
the volume, as true only for a perfect. gas. The
assumption that the mutual attraction of the molecules is inversely proportional to the square of the
volume was regarded by Clausius only as a close
approximation under certain conditions. Clausius
finally suggested the following equation to express

[1] *Proc. Phys. Soc.*, Lond., p. 233, 1897.
[2] *Phil. Mag.* [5], 9, p. 393.

the relation of the pressure and volume to the temperature of a gas:—

$$p = R\frac{T}{v-a} - \frac{c}{T(v-\beta)^2},$$

in which R, c, a, and β are constants. In some respects this equation is a closer approximation than that of Van der Waals. In 1882 Sarrau[1] investigated this equation very thoroughly with reference to the experimental results obtained by Amagat. He found that, in the case of hydrogen and nitrogen, the equation expresses very closely the actual relation of the pressure and volume to the temperature. With certain other gases, however, some deviations were observed.

From the experimental results it is evident that the equations of Van der Waals and Clausius, as well as similar equations of other experimenters, can be regarded only as close approximations.[2] Inasmuch as all of these equations are based upon the kinetic theory of gases, it may be remarked that within recent years considerable opposition has been offered against this theory.

---

[1] *Compt. rend.,* 94, pp. 639, 718, and 845.

[2] As further reference to this subject see, among others, Rankine, *Trans. Roy. Soc.,* 1854, p. 336; Joule and Thomson, *ibid,* 1862, p. 579; Recknagel, *Pogg. Ann.,* 145, p. 469, 1872; Hirn, *Théorie Méchanique de la Chaleur,* 3d ed., II, p. 211; Amagat, *Ann. de Chim. et de Phys.* [5], 28, p. 500, 1883; and *Zeit. für Compr. und flüss. Gase,* 2, p. 178, 1898; Violi, *Phil. Mag.,* 27, p. 527, 1889; J. Traube, *Wied. Ann.,* 61, 2, pp. 380–400, 1897.

CHAPTER IV

LIQUEFACTION OF THE SO-CALLED PERMANENT
GASES

SECTION I

" THE meeting of the French Academy on the twenty-fourth of December, 1877, was a memorable one. On that day the members were told that Cailletet had succeeded in liquefying both oxygen and carbon monoxide at his works at Chatillon-sur-Seine, and that the former gas had also been liquefied by Raoul Pictet in Geneva." The following letter,[1] addressed by Cailletet to Sainte-Claire Deville, was read by Dumas : —

" I have to tell you first, and without losing a moment, that I have just this day liquefied oxygen and carbon monoxide.

" I am, perhaps, doing wrong to say liquefied,.for at the temperature obtained by the evaporation of sulphurous acid, about — 29°, and at a pressure of 300 atmospheres, I see no liquid, but a mist so dense that I infer the presence of a vapor very near its point of liquefaction.

" I write to-day to M. Deleuil for some protoxide of nitrogen, by means of which I shall be able, without doubt, to see oxygen and carbon monoxide flow.

[1] *Compt. rend.*, 85, p. 1217.

"P. S.—I have just made an experiment which sets my mind greatly at ease. I compressed hydrogen to 300 atmospheres, and after cooling down to — 28°, I released it suddenly. There was not a trace of mist in the tube. My gases, carbon monoxide and oxygen, are therefore about to liquefy, as this mist is produced only with vapors which are on the verge of liquefaction. The prediction of M. Berthelot has been completely realized.

" Louis Cailletet.

"December 2, 1877."

Deville added also that Cailletet had successfully repeated his experiments on the condensation of oxygen, on Sunday, December sixteenth, in the laboratory of the Normal School.

The following telegram[1] was then read : —

"Geneva, December 22.

" To-day I liquefied oxygen at a pressure of 320 atmospheres, and a temperature of — 140°, obtained by means of sulphurous and carbonic acids.

"Signed,

" Raoul Pictet."

During the session of the Academy a second telegram[2] was received from Pictet.

" Geneva, December 24, 4.15 P.M.

"A second experiment, performed before numerous assistants, has thoroughly confirmed the results which I communicated to M. Dumas last Saturday.

" Pictet."

---

[1] *Compt. rend.*, 85, p. 1214.   [2] *Ibid.*, p. 1220.

The researches of Andrews had already shown that all gases under the proper conditions of temperature and pressure would pass into the liquid state. The experimental evidence, however, for the so-called permanent gases, was left for these two simultaneous, but independent experimenters. Roscoe and Schorlemmer, in their account of these observations, say: "It is difficult, on reading the descriptions of these experiments, to know which to admire most, the ingenious and well-adapted arrangement of the apparatus employed by Pictet, or the singular simplicity of that used by Cailletet. The latter gentleman is one of the greatest of French ironmasters, whilst the former is largely engaged as a manufacturer of ice-making machinery, and the experience and practical knowledge gained by each in his own profession have materially assisted to bring about one of the most interesting results in the annals of scientific discovery."

EXPERIMENTS OF CAILLETET

Previous to the experiments with oxygen, Cailletet succeeded in liquefying a number of gases. He condensed acetylene[1] to a colorless, extremely mobile liquid, and found its vapor-pressure to be 48 atmospheres at a temperature of 1°, and 103

[1] *Compt. rend.*, 85, p. 851.

atmospheres at 31°.   He also liquefied nitric oxide
and methane.[1]   The former gas condensed at a
temperature of − 11°, when subjected to a pressure of 104 atmospheres.   At a temperature of
+ 8° there was no sign of liquefaction, even

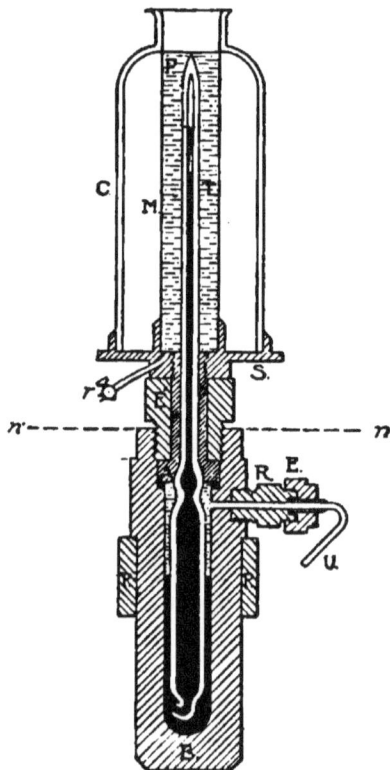

with a pressure of 270
atmospheres.   He con-
cluded that the critical
temperature was be-
tween + 8 and − 11°.
The methane was ob-
tained only in the form
of a mist.   The pure
gas was subjected to
a pressure of 180 at-
mospheres, and a tem-
perature of 7°.   By
releasing the pressure
suddenly, the gas was
condensed to a fine
mist.   These results
were communicated to
the French Academy
by Berthelot, who stated

Fig. 22.

at the same meeting that oxygen and similar gases
could probably be liquefied by the new method
which had been introduced by Cailletet.

[1] *Compt. rend.*, 85, p. 1016.

The apparatus[1] employed by Cailletet in the liquefaction of oxygen and carbon monoxide was similar to that used in the previous experiments. A longitudinal section of the condensing apparatus is shown in figure 22. The glass tube TT is entirely filled with the gas to be compressed. This tube can be removed from the apparatus as shown in figure 23. When the air in the tube has been completely displaced by the gas, the upper end of

FIG. 23.

the tube is hermetically sealed. The lower end is closed with the finger, and the tube introduced into the strong, wrought-iron cylinder B, which is partially filled with mercury. The tube is fastened in the receiver by means of the bronze screw A. The upper portion of the tube is surrounded by a glass cylinder M, containing liquid nitrous oxide, which in turn is surrounded by a safety bell-jar C. n and n' represent gauges for measuring the pressure.

The method of operation is very simple. Water is forced through the tube u into the cylinder B,

[1] *Ann. de Chim. et de Phys.* [5], 15, p. 132.

by means of an hydraulic pump. The mercury
which is displaced by the water passes into the
lower end of the tube *T*. As the pressure increases, the mercury gradually rises, and compresses the gas into a very small space near the
top of the tube.

With this apparatus, Cailletet [1] subjected oxygen
and carbon monoxide to a pressure of 300 atmospheres and a temperature of — 29°. The temperature was reduced in these experiments by
means of liquid sulphurous acid. When the heat,
due to the compression, had been removed from
the gas, and the temperature had become constant, the pressure was suddenly released. In both
instances a mist was formed, showing that a partial condensation had taken place. These observations are really the first experimental evidence of
the liquefaction of the so-called permanent gases.
The condensation of oxygen by Pictet was accomplished about three weeks later. Cailletet's work,
however, was not known to the latter experimenter
at that time. The methods employed by the two
investigators were entirely different; and both
series of observations may be considered as pioneer experiments.

---

[1] *Compt. rend.*, 85, p. 1213.

Liquefaction of Nitrogen

After condensing carbon monoxide and oxygen to the liquid state, Cailletet extended his researches to other gases which had not yet been liquefied. Pure dry nitrogen,[1] when subjected to a pressure of 200 atmospheres and a temperature of 13°, and then suddenly released, formed drops of liquid, of an appreciable volume, which remained in the tube for a period of three seconds. The first experiment was made at a temperature of − 29°. The author says there can be no doubt that the nitrogen was actually liquefied in these experiments.

Liquefaction of Air

After liquefying the gases which compose the atmosphere, Cailletet proceeded to condense ordinary air[2] to the liquid state. He says, " Having liquefied nitrogen and oxygen, the liquefaction of air follows as a matter of course ; but I thought it would be interesting to make a direct experiment, which, of course, succeeded perfectly."

Experiments with Hydrogen

Cailletet also extended his experiments to hydrogen.[3] To obtain evidence of liquefaction with this

---

[1] *Compt. rend.*, 85, p. 1270.   [2] *Ibid.*, p. 1271.   [3] *Ibid.*, p. 1270.

gas was a matter of considerable difficulty. In the first experiment no peculiarities were observed. The observation was repeated many times. When the gas was subjected to a pressure of 280 atmospheres, and then suddenly released, an exceedingly fine and subtle mist was formed throughout the length of the tube. The duration of the phenomenon was extremely short. The experiment was successfully repeated a number of times in the presence of Berthelot, Deville, and Mascart.

EXPERIMENTS OF PICTET [1]

Pictet's work on the liquefaction of oxygen furnishes one of the most brilliant experiments of modern science. It was the culmination of a long experience in the liquefaction of gases and the production of low temperatures. The many difficulties which accompany experiments of this nature were foreseen and provided for. In the introduction to the memoir on these observations, we find that, previous to the experimental part, Pictet made a careful study of the problem. ˙He looked upon cohesion as a property common to the molecules of all forms of matter; hence, he says, all forms of matter can exist in the gaseous, liquid, or solid state. He recognized clearly that press-

[1] *Ann. de Chim. et de Phys.* [5], 13, p. 145; *Archives des Sciences Physiques et Naturelles*, 1878, p. 16

ure alone would not suffice for the condensation of the so-called permanent gases, but he thought that if the temperature could be sufficiently reduced, the molecular cohesion of these gases would condense them into liquids or solids.

The following five conditions were considered by Pictet to be essential in an experiment of this nature : —

1. The gas to be liquefied must be as pure as possible.

2. The experimenter must be able to obtain considerable pressure, and also be able to measure this pressure accurately.

3. The method must be such as to enable the operator not only to obtain a very low temperature, but to maintain this temperature indefinitely.

4. The condensing surface exposed to the low temperature should be as large as possible.

5. Provision should be made for the sudden expansion of the gas which has been subjected to this low temperature and high pressure.

These conditions were all fulfilled in the carefully constructed apparatus of Pictet. It is necessary, in these experiments, that the heat be removed from the gas as rapidly as possible, hence metallic tubes were employed instead of glass.

The apparatus provides for three distinct operations : —

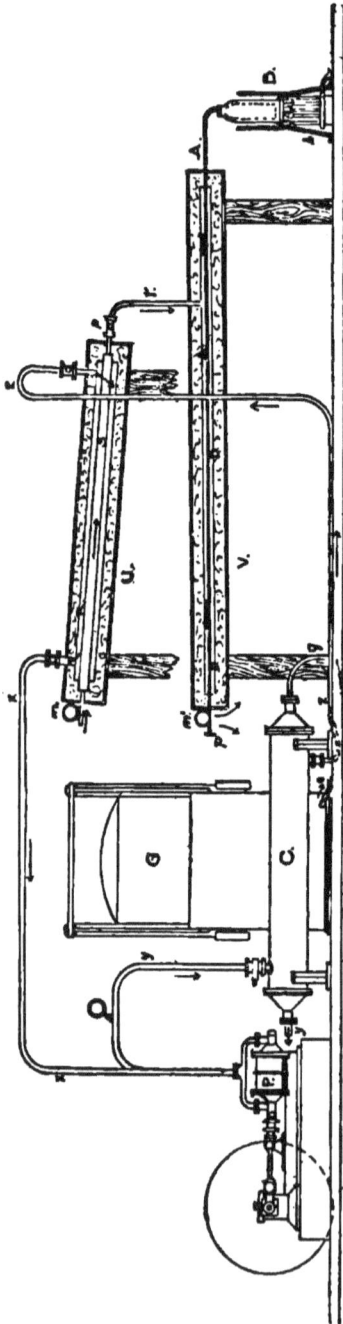

1. The circulation of sulphurous acid, which produces the first lowering of temperature.

2. The circulation of carbonic acid or protoxide of nitrogen, which produces a second lowering of the temperature.

3. The generation of oxygen in a closed vessel provided with a long, narrow tube which is completely surrounded by the carbonic acid.

Figure 24 represents an elevation of this arrangement. U and V are longitudinal sections of two large drums. The former is provided with a large copper tube R, which is about 12 cm. in diameter, and 1.10 m. in length. The tube and drum are slightly

FIG. 24.

inclined, one end being 12 cm. higher than the other.

The liquid sulphurous acid is introduced into this tube through the pipe z, which enters at the lower end, and on the upper side of the tube. At the upper end of the tube R is a stop-cock r which is connected by means of a long tube, 25 mm. in diameter, with the suction of the first pump P.

When the pump is operated, and a partial vacuum is produced in the tube R, the sulphurous acid will begin to evaporate immediately, and the temperature of the liquid will fall rapidly. The more perfect the vacuum, the lower will the temperature fall.

In order to make the process continuous for a period of several hours, it was necessary to make use of a second pump, P' (Fig. 25), and arrange the two so that the suction of P' corresponded to the pressure of P. Experiment showed that, by means of this arrangement, the temperature could be reduced almost 20° farther than with one pump alone. The pumps were made of cast iron. The pistons were hollow, and provided with a circulation of water. The valves were carefully constructed of steel. The speed of the pumps varied from 80 to 100 strokes per minute. The pistons of the two pumps were connected by means of a metallic pipe. P' was connected with the copper

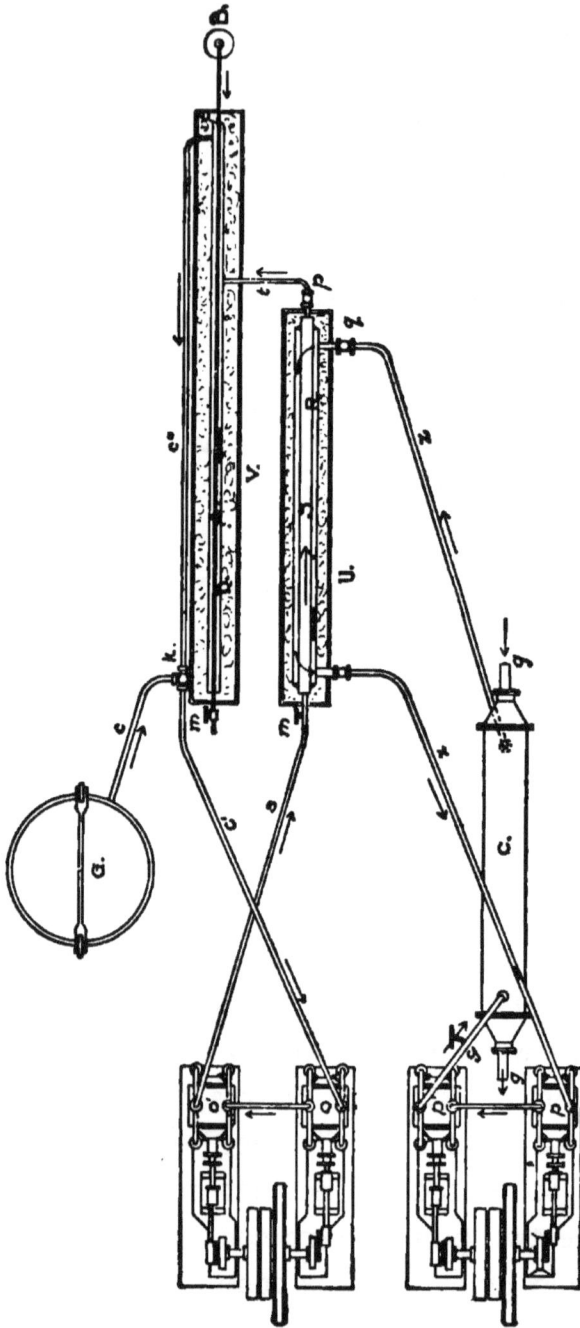

Fig. 45.

condenser C, the tubes of which were traversed by a current of water. The condenser C is also connected with the tube z.

The action is as follows: The sulphurous acid which is evaporated in the tube R passes through the cock r into the first pump P, thence into the second pump P', and finally into the condenser C, where it is compressed to from 1 to 2 atmospheres and condensed to the liquid state. From the condenser C, the sulphurous acid passes through the pipe z, and again enters the tube R. The quantity of acid which passes from c to R can be regulated by means of the screw-valve q, so as to exactly replace the vapor which is exhausted. The quantity of liquid then in the tube R will remain constant.

With this arrangement, the cycle of sulphurous acid is complete; and the fall of temperature produced in the large tube R can be permanently maintained.

The first circulation of sulphurous acid is only an expedient to obtain a sufficient quantity of carbonic acid or protoxide of nitrogen in the liquid state. To accomplish this a tube S, 6 cm. in diameter and 1.15 m. in length, is placed in the tube R. The tube S, of course, is completely immersed in the liquid sulphurous acid, and has a temperature of $-65°$. At this temperature, it requires a pressure of only 4 to 6 atmospheres to condense the carbonic acid.

The carbonic acid was prepared from Carrara marble and hydrochloric acid; it was then carefully washed and dried, and finally stored in the oil gasometer G (Figs. 24 and 25). This vessel was connected, by means of the tube c, with the three-way cock K (Fig. 25). The gaseous carbonic acid was pumped from the vessel G, by means of the pumps O and O', which are similar in every respect to P and P', and compressed in the tube S, where it immediately condensed to the liquid state.

The next step was to utilize the liquid carbonic acid in lowering the temperature of oxygen gas. This was accomplished in the drum V, which is similar to U. The two concentric copper tubes, D and A, were placed in this drum, which was slightly inclined, but in the opposite direction from that of U. The tube D was 3.70 m. in length and 35 mm. in external diameter, while A was 4.16 m. in length and 15 mm. in external diameter. The arrangement is shown in the figures. The tube D is connected with the tube S, which contains the liquid carbonic acid, by means of the pipe t. The flow of acid from S to D is regulated by means of the screw-valve p. The pipe c'' connects the upper part of D with the three-way cock K.

By turning the cock K, the suction of the pump O is brought into connection with the tube D by means of the pipes c' and c''. The pump O' is connected with the tube S by means of the pipe s. With this arrangement the cycle of carbonic acid is complete, just as in the case of sulphurous acid. The directions in which the acids flow at different parts of the apparatus are shown by the arrows. By means of this double circulation a temperature of from -120 to $-140°$ could be obtained.

The tube A is bent downward at one end and connected with the large wrought-iron flask B. This vessel, with a capacity of 1659 cc., was forged with the utmost care, so as to insure homogeneity throughout the metal. The walls of the flask were 35 mm. in thickness. The vessel B contains a definite quantity of a perfectly dry mixture of potassium chlorate and potassium chloride. On heating the flask, by means of the gas burner b, oxygen gas is generated. In this way a pressure of several hundred atmospheres can easily be obtained. The oxygen, of course, is compressed in the tube A as well as in the flask B.

After the apparatus had once been set up, fifteen days were spent in preliminary experiments, in order to test separately the different parts of the apparatus, and to determine the conditions under which the best results could be obtained. A thousand precautions, he says, are necessary to insure the success of the experiment. After all the conditions and details had been carefully studied, Pictet began on the morning of December twenty-second to make a complete experiment. The complete working of the apparatus can be understood from the following details which were given concerning the first experiment: —

9.00 A.M. — The sulphurous acid pumps are started. The temperature in the tube R falls rapidly.

9.30. — The temperature is — 55°. The carbonic acid pumps are started. The gasometer descends. The pressure of the carbonic acid is 6 atmospheres. During the working of the pumps the pressure slowly increases to 8 atmospheres.

9.50. — The temperature is — 49°, the pressure 8.5 atmospheres.

I stop the admission of carbonic acid to the pumps.

10.20. — Temperature — 65°, pressure 3.9 atmospheres.

A slight quantity of gas is again admitted.

10.40. — Temperature — 60°, pressure 5 atmospheres.

800 litres of carbonic acid are now liquefied.

Hoar-frost covers the lower part of the manometer m'.

10.50. — The wrought-iron flask is screwed on to the tube A. It is charged with a mixture of 700 grams of potassium chlorate and 250 grams of potassium chloride powdered together in a mortar, sifted and carefully dried.

11.00. — The gas under the flask is lighted.

Carbonic acid is admitted more freely into the pumps. The pressure increases to 10 atmospheres; the temperature is − 48°. The appearance of frost on the pipe c'' indicates that the carbonic acid has passed over into the long tube D.

11.15. — The pipe c'' is connected with the suction of the pumps. The temperature of the carbonic acid reaches a minimum, − 130°.

11.35. — The manometer m' on the oxygen tube records a pressure of 5 atmospheres.

The circulation of sulphurous and carbonic acids is completely established.

12.10 P.M. — The oxygen manometer records a pressure of 50 atmospheres.

12.16. — The pressure increases to 60 atmospheres, and then gradually rises as follows : —

K

12.23	pressure,	70 atmospheres.
12.29	"	80 "
12.34	"	90 "
12.36	"	100 "
12.37	"	150 "
12.37.25″	"	200 "
12.38	"	460 "
12.39	"	510 "
12.39.30″	"	522 "
12.40	"	525 "
12.42	"	526 "
12.44	"	525 "
12.48	"	505 "
12.50	"	495 "
1.00	"	471 "

1.05. — The pressure is 471 atmospheres.

The pressure is stationary; hence all the chemical and physical phenomena have terminated.

The condensation has produced the fall of pressure recorded by the manometer. The tube A is filled with liquid oxygen.

There is evidently an excess of gas, which causes a higher pressure than that corresponding to the temperature of the liquid carbonic acid.

1.10. — The pressure is exactly 470 atmospheres. The plug which closes the tube A is opened.

A liquid jet issues with great violence, and

assumes the appearance of a brilliant white pencil. A bluish halo surrounds the jet, especially the lower part. The jet of liquid is from 10 to 12 cm. in length, and about 1.5 to 2 cm. in diameter. It continues for a period of 3 or 4 seconds.

The regulating cock is closed. The pressure is 396 atmospheres, but falls rapidly to 352 atmospheres, where it remains stationary for about 3 minutes.

1.18. — The tube is again opened; a second liquid jet, similar to first, issues. But following this, the gas escapes with a characteristic aëriform appearance.

The gas, in expanding, produces a mist by partial condensation. The appearance, however, is quite different from that of the first jet. This indicates an absence of liquid in the tube.

1.19. — The pressure is 50 atmospheres. The gas escapes as a bluish mist, but there is no evidence of liquid being carried with it.

Live coals, when placed under the second jet, blaze up instantly with great violence.

This experiment demonstrated clearly that oxygen can be condensed to the liquid state. Pictet calls special attention to the difference between the mist produced by the expansion of the gas alone, and that produced when there was liquid in the tube. The change, he says, from one to the other was so apparent that it was noticed by more than twenty spectators at the same moment.

Five experiments were made, three with carbonic acid, and two with protoxide of nitrogen. The same quantity of potassium chlorate and potassium chloride was used in each of these experiments.

The more important observations refer to five successive stages, which the author designates as follows : —

1. *The maximum stationary pressure of the oxygen before the first jet is allowed to escape.*

This pressure is always stationary for at least a quarter of an hour, and is always less than the pressure recorded at the end of the chemical reaction.

2. *The pressure immediately after the escape of the first jet*, when it is *distinctly* observed that the liquid has been replaced by a gaseous jet.

3. *The stationary pressure after the first jet.*

The fall of pressure in this operation is due to the condensation of oxygen.

When the tube is filled a second time, condensation ceases, and the pressure becomes stationary.

4. *The pressure immediately after the second jet.*

5. *The stationary pressure after the second jet.*

The third jet was never complete ; it was always

shorter than the first two. This indicated that the condensation was not sufficient to fill the tube three times.

These pressures for five different experiments on the liquefaction of oxygen are shown in the following table. The pressures are given in atmospheres : —

	NUMBER OF EXPERIMENT				
	1	2	3	4	5
1. Maximum stationary pressure before the first jet	470	471	471	469	469
2. Pressure immediately after the first jet	367	395	432	400	416
3. Stationary pressure after the first jet	308	339	378	346	361
4. Pressure immediately after the second jet	285	290	291	285	296
5. Stationary pressure after the second jet	274	271	272	251	253
6. Pressure immediately after the third jet		245		215	205
7. Stationary pressure after the third jet		253		218	212
8. Pressure after the fourth jet. In this case the jet was entirely gaseous	0	0	0	0	0

Pictet endeavored to determine the density of liquid oxygen. The method, of course, was

indirect, and involved numerous calculations.   In
the original memoir, several pages are devoted to
this part of the experiment.   He concluded that
the density of liquid oxygen was very nearly
equal to that of water.   Later researches have
shown the density to be slightly greater than
that of water, but considering the data upon
which Pictet based his calculations, the result is
remarkably accurate.

Measurements on the vapor-pressure of liquid
oxygen showed it to be about 270 atmospheres at
the temperature of the liquid carbonic acid, and
about 250 atmospheres at the temperature obtained
by means of the protoxide of nitrogen.

Solidification of Oxygen

In the third and fourth experiments the escaping jet was carefully examined by means of an
electric light.   The jet appeared to consist of two
parts; a central portion of 2 to 3 mm. in diameter,
and an outer envelope.   It resembled two concentric cylinders, the inner of which was partially
transparent, while the outer appeared as snow-
white dust.   The light reflected, by the dust, at
right angles to the incident ray was examined
by means of a polariscope, and found to be partially polarized.   This examination was made by

M. H. Dufour, Professor of Physics in the Academy of Lausanne. Although the experiment was not conclusive evidence, it was thought by those present that the outer portion of the jet consisted of solid particles of oxygen.

Experiments with Hydrogen

"Having obtained the preceding results with oxygen," Pictet says, "we were naturally induced to treat hydrogen in a similar manner. The entire mechanical apparatus used for the first gas can be used for the second without alteration."

In this experiment the hydrogen was generated from 1261 grams of potassium formate and 500 grams of caustic potash. The two were mixed. together, and carefully dried.

The temperature was lowered by means of sulphurous acid and protoxide of nitrogen. The method of procedure was exactly the same as in the case of oxygen.

The following are some of the notes recorded in the first experiment, which was begun at 7 P.M. Jan. 10, 1878 : —

The pumps were started at 7 o'clock, and when the circulation of the two acids was complete, the gas under the hydrogen generator was lighted.

8.32   P.M. — The pressure of hydrogen was   50 atmospheres.
8.47   "   "   "   100   "
9.00   "   "   "   200   "
9.04   "   "   "   300   "
9.06.15″   "   "   "   400   "
9.07   "   "   "   500   "
9.08   "   "   "   550   "
9.10.30″   "   "   "   640   "
9.11   "   "   "   650   "
9.11.30″   "   "   "   652   "

At this point the pressure became almost stationary, and the tube containing the hydrogen was opened. An opaque jet of a bluish tint issued from the orifice. The jet became intermittent, and Pictet concluded that the hydrogen had solidified at the opening in the tube.

From the observations recorded in this experiment, there can be but little doubt that, at the moment when the pressure was relieved, the hydrogen was partially liquefied and probably solidified.

Cailletet and Pictet removed the last doubt as to the possibility of liquefying such gases as oxygen, nitrogen, and hydrogen. The idea of permanent gases is no longer tenable. Since these experiments were made, the apparatus for the liquefaction of gases has been materially modified and improved. The names of these two investigators, while connected especially with the pioneer work in the liquefaction of the permanent gases, are associated also with the names of the more recent experimenters.

LIQUEFACTION OF OZONE BY HAUTEFEUILLE AND CHAPPUIS

In 1882 Hautefeuille and Chappuis condensed ozone to the liquid state.[1] The gas was compressed in the Cailletet apparatus to a pressure of 125 atmospheres. The temperature was lowered by means of a jet of liquid ethylene to about — 100°. Under these conditions the gas condensed to an indigo-blue liquid. The liquid remained for a short time in a static condition at the ordinary atmospheric pressure. The authors state that the ozone probably contained some oxygen. This gas has been liquefied by later experimenters, who have also determined its critical constants and boiling point.

SECTION II

EXPERIMENTS OF WROBLEWSKI AND OLSZEWSKI

In 1883 the names of two brilliant scientists, Wroblewski and Olszewski, were added to the list

---

[1] *Compt. rend.*, 94, p. 1249.

of experimenters on the liquefaction of gases. These two observers, co-workers at times, and independent investigators at other times, have done much to bring about the present state of perfection in the methods employed in the condensation of gases.

Their first work in this direction consisted in the liquefaction of oxygen, nitrogen, and carbon monoxide.[1] Acting upon the suggestion of Cailletet they employed liquid ethylene as a refrigerant.

The apparatus employed is represented in figures 26 and 27. The gas was compressed by means of the contrivance[2] shown in figure 26. The gas is introduced into the large glass tube i which rests in the hollow iron cylinder a b. The upper end of the cylinder is closed air-tight by means of the brass contrivance d, which is held in its position by means of the brass screw c. A small steel tube extends from the upper end of the tube i throughout the brass portion d. The side tube p serves to connect the interior of the apparatus with a manometer and a Cailletet pump. The method of operation is similar to that of Cailletet. The large tube i and the smaller tube

---

[1] *Wied. Ann.*, 20, p. 243, 1883.

[2] This portion of the apparatus was constructed by Wroblewski in 1882, for the purpose of studying surface tension. *Compt. rend.*, 95, pp. 284 and 342.

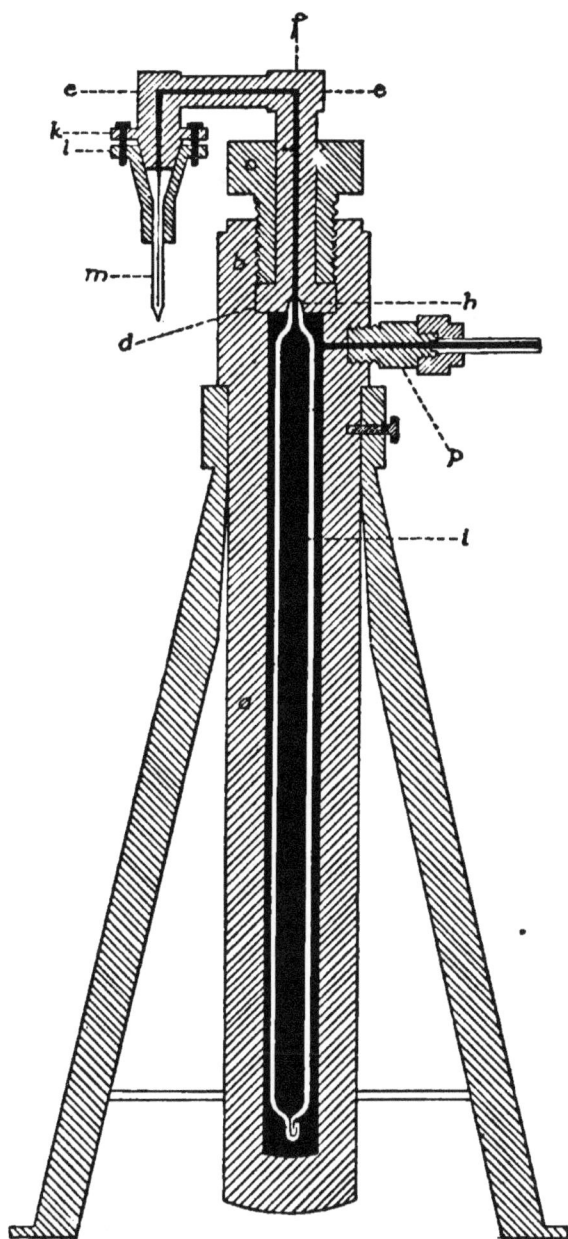

FIG. 26.

above are filled with the gas to be compressed. Mercury is then introduced through p until the desired pressure is obtained.   The resisting strength of the apparatus was estimated to be about 500 atmospheres.

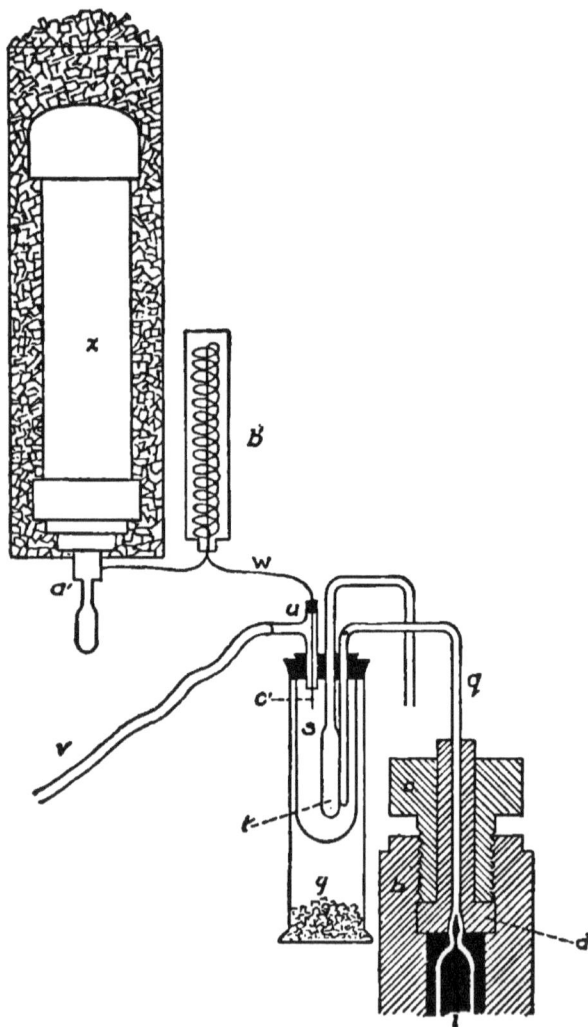

FIG. 27.

The liquefying apparatus is represented in figure 27. The compression apparatus is shown in the lower right-hand corner. In this case, however, the steel tube d is replaced by the thick-walled capillary glass tube q which is fused into the upper end of the tube i. The tube i, together with the small tube q, has a capacity of about 200 cubic centimetres. These tubes are filled with the gas to be liquefied by means of a Jolly mercury pump. The tube q passes air-tight into the glass vessel s which, in turn, is closed air-tight by means of a rubber stopper. t is a hydrogen thermometer constructed on the principle of the Jolly air thermometer. The copper tube w, which passes through the T-tube u, leads to the liquid ethylene in the receiver x of a Natterer compression pump. This receiver rests in the large zinc vessel z which contains a freezing mixture. The spiral portion of the tube w is surrounded by solid carbonic acid and ether in the vessel b. The side tube of u is connected with the lead tube v, which leads to a Bianchi exhaust pump. The liquid ethylene then can be evaporated under reduced pressure.

Liquefaction of Oxygen

The oxygen gas was prepared from pure potassium perchlorate, washed with caustic potash, and dried by means of concentrated sulphuric acid.

The tubes *i* and *q* were then filled with the gas. The tube *q* was cooled down to a temperature of — 130°. The pressure was then increased to more than twenty atmospheres, when the gas was completely liquefied. The liquid oxygen collected in the lower end of tube *q*, and showed a well-defined meniscus. Measurements were made of the vapor-pressure at different temperatures.

Liquefaction of Nitrogen and Carbon Monoxide

After the successful experiment on the liquefaction of oxygen, the investigation was extended to nitrogen and carbon monoxide. When subjected to a temperature of — 136°, and a pressure of 150 atmospheres, both of these substances remained in the gaseous condition. When the pressure was suddenly released, however, there was evidence of liquefaction in the tube. By relieving the pressure more slowly and keeping it always above 50 atmospheres, the nitrogen and carbon monoxide were both obtained as liquids in a static condition. The liquids were transparent and colorless, and showed, in each case, a well-defined meniscus.

These observations were the beginning of a long series of experiments by these investigators. Their later researches were carried on independently, and will be considered presently.

DEMONSTRATION OF THE LIQUEFACTION OF OXYGEN BY DEWAR

During the next year after the liquefaction of oxygen and nitrogen by Wroblewski and Olszewski, James Dewar of London described an apparatus[1] for demonstrating the liquefaction of oxygen in the lecture room. Since that time he has been a constant worker in the field of the liquefaction of gases, and is the author of many important contributions.

From this point it is probably advisable to consider the work of Wroblewski, Olszewski, and Dewar separately. The historical development of the methods can be gathered from the dates of the experiments.

EXPERIMENTS OF WROBLEWSKI

Soon after the experiments of Wroblewski and Olszewski on the liquefaction of oxygen, air, nitrogen, etc., in 1883, the former experimenter made some important observations in the same field. Early the next year he subjected hydrogen[2] to a pressure of 100 atmospheres in a small glass tube. The temperature of the gas was lowered by the evaporation of liquid oxygen which surrounded the

---

[1] *Proc. Roy. Inst.*, 11, p. 148;   *Phil. Mag.*, 18, p. 210, 1884.
[2] *Compt. rend.*, 98, p. 149, 1884.

liquefying tube.   On suddenly releasing the pressure, signs of ebullition were observed in the tube,
showing that the hydrogen had been partially
liquefied.

In April of the same year Wroblewski published
an article [1] on the boiling points of oxygen, air,
nitrogen, and carbon monoxide. The boiling points
of oxygen under different pressures were as
follows : —

Pressure	Boiling Point
50 atmospheres	$- 113°$
27.02 "	$- 129.6$
24.40 "	$- 133.4$
22.20 "	$- 135.8$
1. "	$- 184.$

The last temperature represents the boiling
point of oxygen under the ordinary atmospheric
pressure. The following values were obtained for
the boiling points of air, carbon monoxide, and
nitrogen under a pressure of one atmosphere : —

Substance	Boiling Point
Carbon monoxide	$- 186°$
Air	$- 192.2$
Nitrogen	$- 194.3$

[1] *Compt. rend.*, 98, p. 982, 1884.

By evaporating liquid air and nitrogen under reduced pressure, a temperature of $-200°$ was obtained. The author stated in this article that atmospheric air will be the refrigerant of the future.

A few months later the same experimenter[1] liquefied methane and studied its properties. The critical temperature he found to be $-73°.5$, and the critical pressure 56.8 atmospheres. The boiling point, he says, is between $-155°$ and $-160°$. By using liquid marsh-gas as a refrigerant,[2] the author states that oxygen, air, carbon monoxide, and nitrogen, may be liquefied at comparatively low pressures.

In 1885 Wroblewski published a detailed account of the apparatus[3] employed by him in the liquefaction of oxygen, carbon monoxide, air, nitrogen, etc. The plan of the apparatus is shown in figure 28.

The two iron cylinders, a and b, previously tested to 150 atmospheres pressure, are provided at the ends with the screw-cocks c, d, e, and f. The two vessels are connected with each other and with the manometer h by means of the cop-

[1] *Compt. rend.*, 99, p. 136, 1884.

[2] The use of liquid methane as a refrigerant was first suggested by Dewar, *Nature*, 28, p. 551, 1883.

[3] *Wien. Ber.*, 91, p. 672, 1885; *Wied. Ann.* 25, p. 371, 1885.

L

per tube *gg*. This connection, however, can be interrupted at any time by means of the screwcocks at the ends of the vessels. The cylinders are fastened by means of the metal strip *i* in the

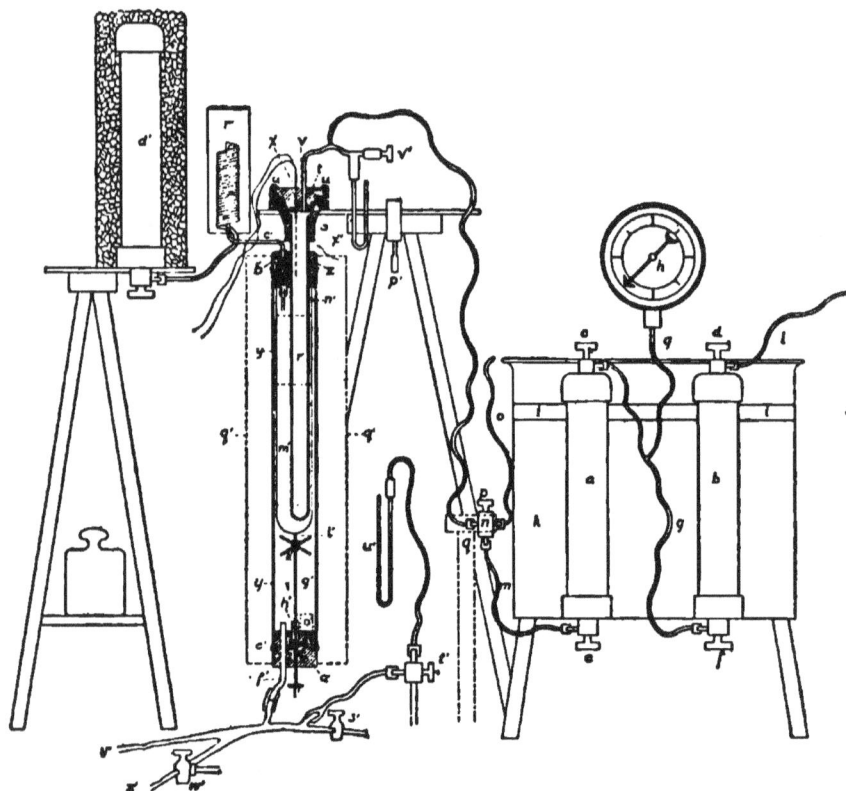

FIG. 28.

zinc chamber *k*, which can be filled, if desirable, with a freezing mixture. The copper tube *l*, about three metres in length, is connected with a Natterer pump. The screw-cock *e* is connected by means of the tube *m* with the steel contrivance *n*, which in turn is connected with the liquefying apparatus and the air manometer *o*. This contrivance is held firmly in its position by means of the iron frame *q*. The pressure in the vessel *a* is increased to the liquefying pressure, not less than 40 atmospheres, while in *b* it ranges from 100 to 120 atmospheres. Each of the vessels, *a* and *b*, contain small cylinders of calcium chloride and caustic potash, to remove any moisture and carbon dioxide from the gas.

The gas is liquefied in the thick-walled glass tube *r*, 42 to 46 cm. in length, which is capable of resisting a pressure of 60 atmospheres. The lower end of the tube is sealed, while the upper end is cemented into the brass flange *s*, and is closed airtight by means of the contrivance *t*. The opening *v* is connected by means of the copper tube *w* with *n*, which in turn is connected with the compression cylinders. The opening *x* serves for the introduction of a thermometer which, in this case, was a thermo-electric pair connected with a sensitive reflecting galvanometer. The tube *x* is placed in a larger tube *m'*, which is connected with the cylinder *d'* containing liquid ethylene. The tube *m'*, in turn, rests in the larger tube *y*, and communicates with it at the small opening *n'*. The vessel *r'* contains solid carbonic acid and ether, through which the liquid ethylene must pass before it reaches the liquefying tube. The ethylene vapors pass out of the apparatus through the tube f', which is connected with an air-pump to reduce the pressure. At a pressure of about 10 mm. of mercury the temperature sank to $-152°$. When the temperature is thus lowered, the pressure in the liquefying tube is increased to about 40 atmospheres, when the gas condenses to the liquid state Considerable care is required to get the apparatus filled with perfectly dry gas.

In this way Wroblewski liquefied oxygen in considerable quantity, and kept it in a static condition, at the ordinary atmospheric pressure, for a period of a quarter of an hour. He also liquefied nitrogen, air, and carbon monoxide, and studied their properties. The following values were obtained by him for the critical constants of these gases : —

Substance	Critical Temperature	Critical Pressure
Oxygen	-118	50 atmospheres
Carbon monoxide . .	-140.2	39 "
Nitrogen	-145.5	35 "

The author also made a careful study of the measurement of low temperatures. For this purpose he made use of a copper-German-silver

thermo-electric pair and a sensitive reflecting galvanometer. Previous to the measurements, the
apparatus was compared with a hydrogen thermometer. By the evaporation of oxygen at different pressures, the following temperatures were
obtained : —

Pressure	Temperature
74 cm. of mercury	− 181°.5
16 " "	− 190.0
10 " "	− 190 5
8 " "	− 191.98
6 " "	− 194.4
4 " "	− 197.7
2 " "	− 200.4

The oxygen in these experiments was partially
solidified.

The following temperatures were obtained by
the evaporation of carbon monoxide : —

Pressure (in cm. of mercury) :

    73.5   16   10   6   4

Temperature :

  −190° −197°.5 −198°.83 −201°.5 −201°.6

The carbon monoxide was completely solidified.

The temperatures obtained by the evaporation of
liquid nitrogen are as follows : —

Pressures :   74   12   8   6   4.2

Temperatures : −193° −201° −201°.7 −204° −206'

The nitrogen at the lowest temperature was partially solidified.

The author also determined the vapor-pressures of liquid oxygen, carbon monoxide, and nitrogen at different temperatures.

Liquefaction of Air

In 1885 Wroblewski made an elaborate series of experiments on the liquefaction of air.[1] The apparatus employed was similar to that represented in figure 28. The object of these experiments was to separate liquid air into two distinct layers.

In a preliminary experiment, a mixture of five parts of carbonic acid and one part of air, at a temperature of 0°, was compressed until a portion of the gas was condensed to the liquid state. In this condition the gas and liquid in the tube were separated by a well-defined meniscus. As the pressure was increased, the meniscus became less distinct, and finally disappeared at the instant when the optical density of the gas became equal to that of the liquid. By increasing the pressure still further, a second meniscus was produced, which occupied a much higher position than the first. In this condition the tube contained two liquids, separated by a sharp meniscus. After a

[1] *Wied. Ann.*, 26, p. 134.

time the liquids mixed completely, and formed a homogeneous liquid.

In a similar manner liquid atmospheric air was obtained in two distinct layers, separated by a well-defined meniscus. The air was partially liquefied at a temperature of − 142°, which is above the critical temperature of nitrogen. When the pressure was increased to 40 atmospheres, the meniscus which separated the liquid from the gas disappeared. The temperature was then lowered, and, at a pressure of 37.8 atmospheres, a second meniscus was formed, which occupied a higher position than the first. The air in this condition consisted of two layers of liquid separated by a sharp meniscus. Analysis showed that the lower layer contained about three per cent more oxygen than the upper layer. After a couple of minutes the two liquids mixed and formed a homogeneous fluid.

The author also made a large number of observations on the vapor-pressure of liquid air at different temperatures. He also studied carefully the changes of temperature and pressure which accompanied the mixture of the two layers of liquid air.

In addition to the experiments which have been outlined, Wroblewski made numerous observations on the properties of matter at low temperatures.

He also made a study of the relation between the gaseous and liquid states of matter.[1] At the early age of forty years, just in the midst of his active career, he met with a fatal accident. "While working late at night in his laboratory, he fell asleep, and in his sleep he overthrew a kerosene lamp. His clothing began to burn, and the wounds thus received resulted, four days later, in death." "The value of Wroblewski's supplement to the work of Cailletet and Pictet cannot be easily estimated, and we deplore the occurrence of the fatal accident which robs the scientific world of such an able experimenter in this important department of chemistry."[2]

EXPERIMENTS OF OLSZEWSKI

During the first few years of investigation in the field of the liquefaction of gases, Olszewski's experiments were somewhat similar to those of Wroblewski. The work of these two observers was carried on simultaneously until the death of the latter in 1888.

Soon after the liquefaction of oxygen, nitrogen,

---

[1] *Wied. Ann.*, 29, p. 428, 1886.

[2] Wroblewski's work on the liquefaction of gases has been published in the form of a pamphlet entitled, *Comment l'air a été Liquefié.* — Ref. Dewar, *Chem. News*, 73, p. 42.

etc., by Wroblewski and Olszewski in 1883, the latter experimenter made use of liquid oxygen as a refrigerant in the liquefaction of other gases. In 1884 he subjected hydrogen [1] to a pressure of nearly 200 atmospheres, and at the same time cooled the gas to the temperature of liquid oxygen (boiling under a pressure of 6 mm. of mercury). Under these conditions, hydrogen showed no meniscus, but when the pressure was suddenly released a momentary ebullition was observed. The author concluded that, at the temperature obtained by the evaporation of liquid oxygen under reduced pressure, it is impossible to obtain liquid hydrogen in a static condition.

The experiments on hydrogen were repeated during the same year, with similar results.[2] In this series of experiments the author determined the critical pressure of nitrogen, and found it to be 39.2 atmospheres.

Olszewski also made a large series of experiments on the production of low temperatures by the evaporation of liquid oxygen, nitrogen, air, etc. under reduced pressure. The apparatus[3] employed by him in the earlier experiments on the liquefaction of gases is represented in figure 29.

The gas is liquefied in the thick-walled glass tube

---

[1] *Compt. rend.*, 98, p. 365.   [2] *Ibid.*, p. 913.
[3] *Wied. Ann.*, 31, p. 58.

a, which is 30 centimetres in length, and 14 millimetres in diameter. The lower end of this tube

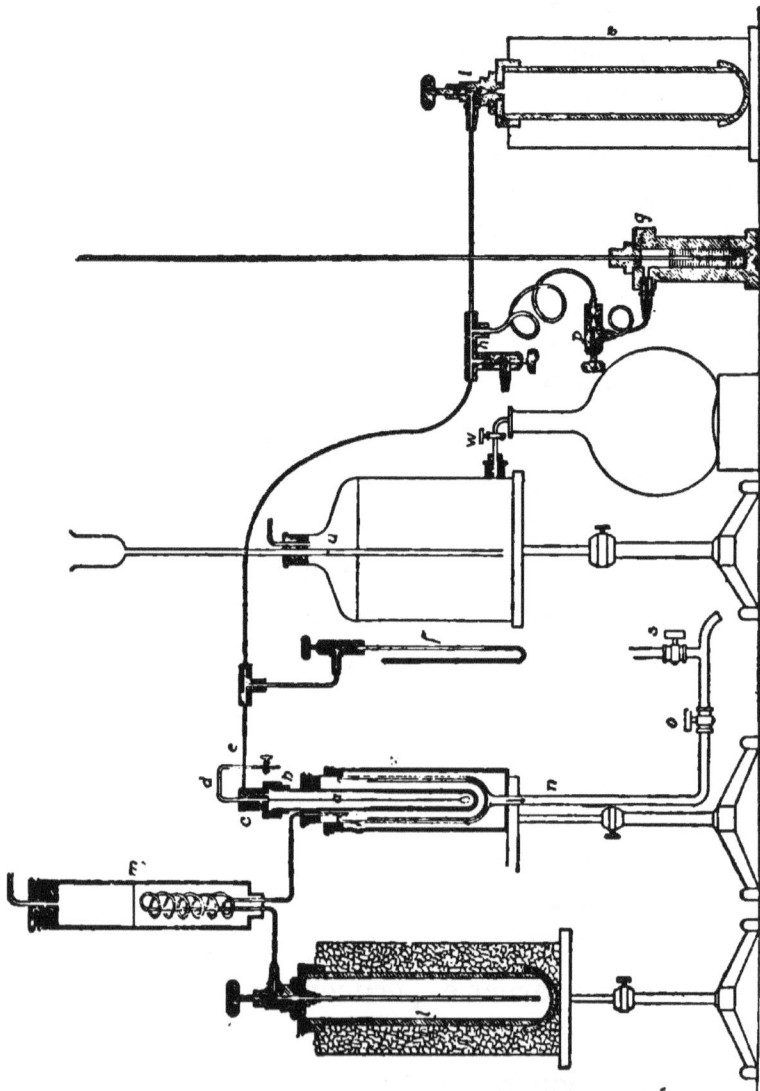

FIG. 29.

is sealed off, while the upper end is slightly wi-
dened, and fastened in the brass flange *b*, which

screws into the brass contrivance c. The tube d
represents a hydrogen thermometer. The liquefying tube a is connected, by means of the copper
tube e, 1st, with the manometer f, which serves for
the measurement of high pressures lower than 1 atmosphere; 2d, with the air manometer g for the
measurement of high pressures; 3d, with a vacuum-
pump; 4th, with the aspirator r; 5th, with the
Natterer receiver i, by means of which a pressure
of 60 to 80 atmospheres may be obtained. The
tube a is placed in a system of concentric glass
tubes, which communicate with each other at the
top. The outer tube is connected with an exhaust-
pump by means of the pipe n. The inner tube of
the system is connected by means of a copper tube
with the Natterer receiver l, which contains liquid
ethylene. The temperature of the liquefying tube
is lowered by the evaporation of this liquid under
reduced pressure. The liquid ethylene, before
entering the system of tubes, passes through a copper coil, surrounded by ether and solid carbonic
acid, in the chamber m. By the evaporation of the
liquid ethylene, under a pressure of 10 mm. of
mercury, the temperature sank below − 150°.

The working of the apparatus is very simple.
After the temperature has been lowered to about
− 150°, the gas is compressed in the tube a by
means of the vessel i, which contains the gas

under high pressure. The condensation to liquid begins in a very short time.

The experiments which have been carried out with this apparatus have been published in numerous articles. In 1884[1] a series of observations were made on the critical temperature and pressure of nitrogen, and on the temperature obtained by the evaporation of liquid ethylene and nitrogen under reduced pressure. The results obtained with ethylene are as follows : —

Pressure	Temperature
750 mm. of mercury	− 103°
107 " "	− 115.5
31 " "	− 139
9.8 " "	− 150.4

The temperatures obtained by the evaporation of liquid nitrogen are : —

Pressure	Temperature
35 atmospheres	− 146°
17 "	− 160
1 "	− 194.4 (boiling point)
In vacuo	− 213

[1] *Compt. rend.*, 99, p. 133.

A few months later he made use of liquid air[1] as a refrigerant, and measured the temperatures obtained under different pressures.

Pressure	Temperature
39 atmospheres	− 140° (critical point)
14 "	− 146
4 "	− 176
1 "	− 191.4 (boiling point)
In vacuo	− 205

Solidification of Gases

In carrying out the experiments at low temperatures, Olszewski succeeded in solidifying a number of gases, and in measuring their melting points. In 1884[2] he evaporated liquid carbon monoxide under reduced pressure, and obtained a very low temperature. At a pressure of one atmosphere the temperature was − 190°; this represents the boiling point of carbon monoxide. By reducing the pressure as far as possible the temperature sank to − 211°, and the substance solidified to an opaque mass.

Early the next year a series of similar experiments were made with nitrogen.[3] At a tempera-

---

[1] *Compt. rend.*, 99, p. 184.   [2] *Ibid.*, p. 706.
[3] *Ibid.*, 100, p. 350.

ture of $-214°$ the substance solidified to an opaque mass. By evaporating the resulting solid under a pressure of 4 mm. of mercury the temperature sank to $-225°$.

During the same year methane and nitric oxide were evaporated under reduced pressure.[1] The former gas solidified at a temperature of $-185°.8$. At this temperature the vapor-pressure was 80 mm. of mercury. When the pressure was reduced to 5 mm. the temperature sank to $-201°.5$. Nitric oxide solidified at a temperature of $-167°$. Under a pressure of 18 mm. of mercury a temperature of $-176°.5$ was obtained.

Substance	Freezing Point
Chlorine	$-102°$
Hydrochloric acid	-116
Hydrofluoric acid	-92.3
Phosphine	-133
Arsine	-119
Stibine	-91.5
Ethylene	-169
Silicon tetrafluoride [2]	-102
Hydrogen selenide [8]	-68

[1] *Compt. rend.*, 100, p. 940.
[2] Silicon tetrafluoride does not melt, but sublimes into a gas at the ordinary pressure.
[8] *Phil. Mag.*, 39, p. 210,

Olszewski also solidified a number of other gases[1] some of which had been liquefied many years before. The freezing points are given in the preceding table.

Ethane and Propane remained liquid at a temperature of − 151°.

In 1885 Olszewski made use of liquid air as a refrigerant.[2]   By evaporating this liquid under a very low pressure, a temperature of − 220° was obtained.   An account is also given in this article of the condensation of a mixture of two volumes of hydrogen with one volume of oxygen to a colorless liquid which formed in thin layers.

Densities and Boiling Points of Liquid Methane, Nitrogen, and Oxygen

In 1886 methane, oxygen, and nitrogen[3] were liquefied by means of the apparatus described on page 154. A series of observations were then made on the densities of these liquids at their boiling points. The following results were obtained : —

Liquid	Density
Methane	0.415 (mean of 3 determinations)
Oxygen	1.124 (" 6 ")
Nitrogen	0.885 (" 4 ")

---

[1] *Wien. Monatshefte für Chem.*, 5, p. 127; *Wien. Ber.*, 94, p. 209.
[2] *Compt. rend.*, 101, p. 238.   [3] *Wied. Ann.*, 31, p. 58.

The following boiling points were also determined : —

Liquid	Boiling Point
Methane	$-164°$
Oxygen	-181.4
Nitrogen	-194.4
Carbon monoxide	-190
Nitric oxide	-153

These temperatures were measured by means of a hydrogen thermometer.

Liquefaction of Ozone

Olszewski repeated the experiments of Hautefeuille and Chappuis on the liquefaction of ozone, with a view of determining the boiling point.[1]   The ozone was prepared by means of Siemens' apparatus.   In order to obtain the liquid in a static condition under a pressure of one atmosphere, liquid oxygen was used as the refrigerant.   The apparatus previously described was employed in this experiment.

At the temperature of boiling oxygen ($-181°.4$), the ozone condensed to a dark blue liquid.   In layers of two millimetres in thickness the liquid

[1] *Wied. Ann.*, 37, p. 337.

appeared opaque. The liquid oxygen which surrounded the tube of ozone was evaporated under
reduced pressure with a view of solidifying the
ozone. The experiment, however, was unsuccessful. The boiling point was determined, and found
to be − 106°. The liquid is somewhat explosive,
especially when brought into contact with liquid
ethylene.

*Apparatus for the Liquefaction of Gases on a
Large Scale*

According to the earlier observations of Olszewski it appears that liquid oxygen is a better refrigerant than liquid air or nitrogen. In order that
this liquid might be used as a refrigerant in the
liquefaction of other gases, the author constructed
an apparatus by means of which oxygen can be
liquefied on a comparatively large scale. This
apparatus was constructed in 1890, and enlarged
during the same year.[1]   A section of the enlarged
apparatus is shown in figure 30.

The gas is liquefied in the steel cylinder *a*, of
about 200 cc. capacity. The upper end of this cylinder is connected by means of a small copper
tube with a metallic manometer *b*, and an iron
vessel *c* of 10 litres capacity. The vessel *c* con-

tains dry air or oxygen under a pressure of 100 atmospheres. The lower end of the cylinder *a* is connected by means of a copper tube with the screw-cock *d*, which serves as an outlet for the

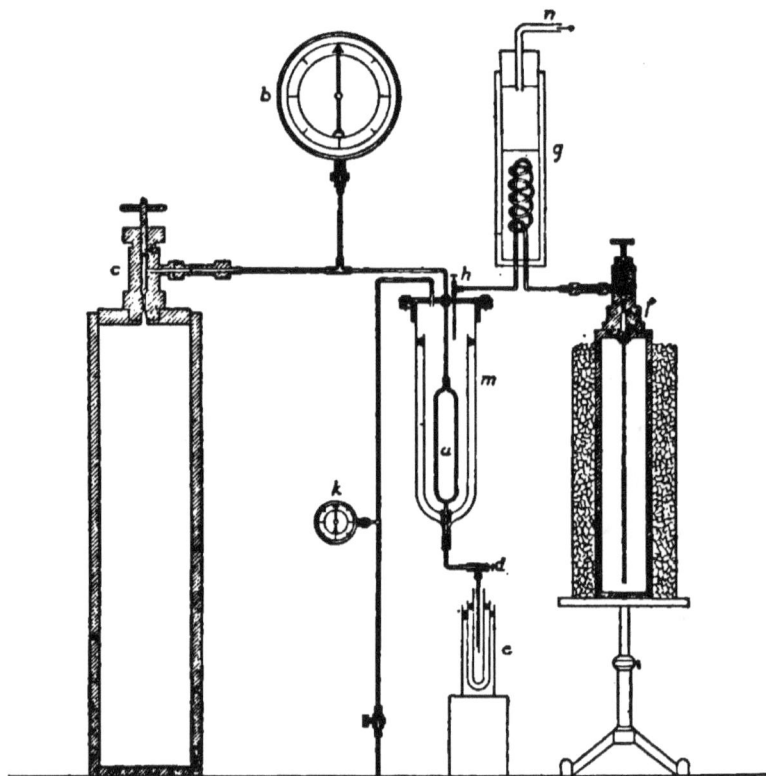

FIG. 30.

liquid oxygen or air. The double- or triple-walled glass vessel *m* which surrounds the cylinder *a* is connected with the reservoir *f*, which contains liquid ethylene, and with an exhaust-pump, by means of the tube *i*. The liquid ethylene, before entering the vessel *m*, passes through the chamber *g*, which contains a mixture of solid carbonic acid and ether. The temperature of this mixture is lowered still further by connecting the tube *n* with an exhaust-pump. The method of operation is as follows : —

The pressure in the chamber *g* is first lowered to 50 mm. of mercury. The vessel *m* is then brought into communication with the exhaust-pump and the reservoir of liquid ethylene. The ethylene evaporates rapidly at first, but finally collects as a liquid in the vessel *m*. As soon as the liquid ethylene completely surrounds the vessel *a*, the supply is cut off. When the temperature is lowered to the critical temperature of the gas contained in the vessel *c*, communication is established between this vessel and the cylinder *a*. The gas enters the cooled cylinder under a pressure indicated by the manometer *b*, and soon begins to liquefy. The manometer, during the condensation, shows a constant fall, and becomes stationary only when the cylinder *a* is completely filled with liquid. When this is accomplished the vessel *c* is closed, and the liquid oxygen or air is allowed to pass out through the cock *d* into the double- or triple-walled glass vessel *e*, after which the process may be repeated.

The temperatures in these experiments were

not measured directly, but were calculated from
the pressure under which the liquid ethylene was
allowed to evaporate. This pressure was measured by means of the metallic vacuometer k.

With this apparatus Olszewski obtained 200
cubic centimetres of liquid air, and even larger
quantities of liquid oxygen. In the earlier experiments it was thought that the liquid oxygen was
colorless, but when obtained in a larger quantity it
was found to be pale blue in color.

The Critical Constants and Boiling Point of Hydrogen

In 1891 Olszewski[1] made another series of
experiments with hydrogen. The apparatus employed was a modification of that represented in
figure 30. Liquid oxygen and liquid air were
used as refrigerants in these experiments. The
author was unable to obtain any indications of a
meniscus of liquid hydrogen, but on allowing the
gas to expand at very low temperatures signs of
ebullition were observed. Moreover, when the
hydrogen was allowed to expand slowly from
pressures of 80, 90, 100, 110, 120, and 140 atmospheres, the phenomenon of ebullition always
appeared at a pressure of 20 atmospheres. From

---

[1] *Phil. Mag.* [5], 39, p. 199.

these observations Olszewski concluded that 20 atmospheres represent the critical pressure of hydrogen.

In order to test the accuracy of this method of determining the critical pressure of a gas, the experiments were extended to oxygen and ethylene. The critical pressure of oxygen had already been measured, and found to be 50.8 atmospheres. To determine the same constant by the expansion method, the gas was cooled to a temperature of about 16°.3 above the critical point, and then allowed to slowly expand. In every case the meniscus and signs of ebullition appeared at a pressure of about 51 atmospheres, provided the initial pressure was not lower than 80 atmospheres. The author concluded, from his experiments on oxygen and ethylene, that critical pressures can be determined very closely by means of the expansion method.

A few years later [1] the author endeavored to measure the temperature of the hydrogen at the moment of expansion. For this purpose a very sensitive platinum resistance thermometer was employed. The thermometer was previously compared with a hydrogen thermometer at temperatures varying from 0° to − 208°.5. When this

[1] *Phil. Mag.* [5], 40, p. 202.

comparison had once been made it was a simple matter to calculate temperatures lower than $-208°.5$. The following results were obtained :—

Expansion of Hydrogen from a High Pressure to	Temperature
20 atmospheres (critical pres.)	$-234°.5$ (critical temperature)
10 "	-239.7
1 "	-243.5 (boiling point)

The experiments were then extended to oxygen, which was allowed to expand at a temperature of about $16°$ above the critical point. The critical temperature of oxygen determined in this way was found to be $-118°$ to $-119°.2$, and the boiling point $-181°.4$ to $-182°.5$. The corresponding values measured directly with a hydrogen thermometer are $-118°.8$ and $-181°.4$ to -182.7, respectively.

Olszewski concluded, from these observations, that the critical temperature of hydrogen is about $-234°.5$, and the boiling point $-243°.5$. These values agree very closely with those obtained theoretically by Natanson.[1]

In connection with the experiments which have been outlined, Olszewski also made numerous experiments on the properties of liquefied gases, and

[1] *Phil. Mag.* [5], 40, p. 272, 1895.

on the properties of matter in general at low temperatures. These investigations were carried out, for most part, in collaboration with his colleague Witkowski. The observations of Olszewski on argon and helium will be considered in Section IV of this chapter.

EXPERIMENTS OF DEWAR

Reference has already been made to the experiments of Dewar on the liquefaction of gaseous mixtures, and the determination of their critical constants. These observations were made in 1880. Four years later he extended his experiments to the so-called permanent gases, and constructed an apparatus for demonstrating the liquefaction of oxygen in the lecture room.[1] The essential parts of the apparatus are shown in figure 31. The gas to be liquefied is contained in the iron reservoir C under a pressure of 150 atmospheres. This reservoir is connected with the manometer D, and also, by means of the small copper tube I, with the liquefying tube F. The pressure is regulated by means of the screw-cock A. The liquefying tube is placed in the glass tube G, which contains liquid ethylene, solid carbonic acid, or liquid nitrous oxide. This tube rests in a larger glass

[1] *Proc. Roy. Inst.*, 1884, p. 148 ; *Phil. Mag.* [5], 18, p. 210.

tube, with which it communicates through the
opening E. The outer tube is connected with a
Bianchi exhaust-pump, so that the liquid ethylene,
etc., can be evaporated under reduced pressure.

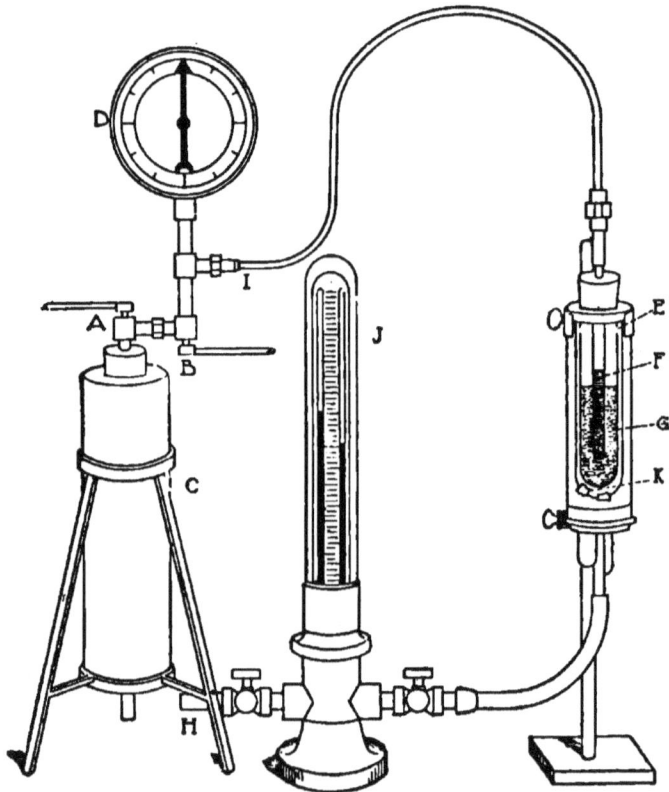

FIG. 31.

The pressure under which the evaporation takes
place is measured by means of the gauge J.

The oxygen, at the temperature obtained by the
evaporation of liquid ethylene under a pressure of
25 millimetres of mercury, could easily be lique-

fied at a pressure of from 20 to 30 atmospheres. When liquid nitrous oxide or solid carbonic acid were used as refrigerants, it was necessary to make use of the sudden expansion of the gas to lower the temperature still further. This was accomplished by increasing the pressure in the liquefying tube to 80 atmospheres, and then opening the screwcock B.

This method was thoroughly satisfactory for demonstrating the liquefaction of oxygen. The liquefying tube was only about five millimetres in diameter, and hence only small quantities of liquid could be obtained. The tube filled with liquid oxygen (for projection) contained about 1.5 cubic centimetres. Some rough observations were also made on the density of liquid oxygen.

In 1886 Dewar[1] published an account of an apparatus by means of which oxygen, air, etc., can be liquefied on a much larger scale. A section of the apparatus is represented in figure 32. The chamber b contains a mixture of solid carbonic acid and ether. Ethylene is conducted into the tube a, where it is liquefied by means of the low temperature produced by the carbonic acid. This liquid passes from the coiled tube into the chamber d, which is surrounded by a larger vessel con-

[1] *Proc. Roy. Inst.*, 1886, p. 550.

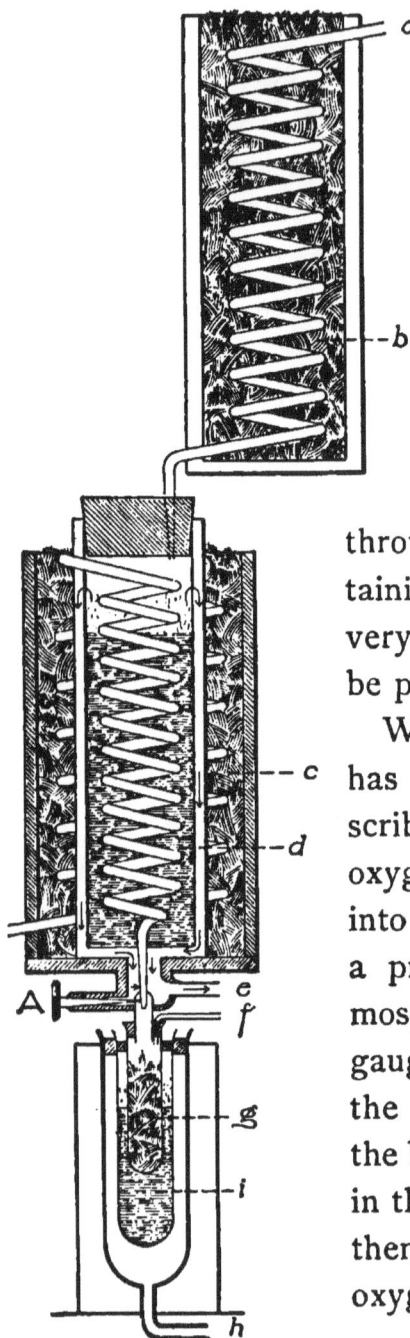

taining solid carbonic acid and ether. The liquid ethylene evaporates into the space between the two chambers. A continuous copper tube, about forty-five feet in length, passes first through the outer vessel, and then through the chamber containing liquid ethylene. A very low temperature can be produced in this way.

When the temperature has been lowered, as described above, perfectly dry oxygen gas is introduced into the copper tube under a pressure of about 75 atmospheres. The pressure gauge (not represented in the figure) soon indicates the beginning of liquefaction in the tube. The valve A is then opened and the liquid oxygen rushes out into the

FIG. 32.

glass tube *g* which is immersed in the liquid ethylene contained in the tube *i*. In the first experiments about 22 cubic centimetres of liquid oxygen were obtained in the tube. In order to reduce the temperature as far as possible, the tubes *g* and *i* were each connected with an exhaust-pump. A temperature of about − 200° was thus obtained. During this process a white deposit was formed in the tube *g*, which was supposed, then, to be solid oxygen. Later, however, the author observed that these white particles of solid matter were due to impurities.

Having constructed an apparatus for liquefying oxygen, air, etc., in considerable quantity, Dewar extended his observations on the properties of liquefied gases, and the properties of matter in general at low temperatures. In 1892 he published an account of some observations on the magnetic properties of liquid oxygen.[1] Becquerel was the first to call attention to this property of oxygen. He experimented with charcoal which was saturated with oxygen gas. In 1849 Faraday noticed that oxygen is strongly magnetic in comparison with other gases. Dewar observed that if liquid oxygen is placed between and a little below the poles of an electro-magnet, the liquid will rise

[1] *Proc. Roy. Inst.*, 13, p. 695.

to the poles when the circuit is completed. The magnetic moment of liquid oxygen, he says, is about 1000, when the magnetic moment of iron is taken as 1,000,000. When liquid air was placed between the poles of the magnet, all of the liquid was drawn to the poles, and there was no separation of oxygen and nitrogen.

The absorption spectrum of liquid oxygen was also examined, and found to be the same as that of the gas. Both the liquid and the highly compressed gas show a series of five absorption bands, situated respectively in the orange, yellow, green, and blue of the spectrum. Some experiments were also made on chemical action at low temperatures. Reference will be made to these observations in the conclusion at the end of the volume.

During the next year Dewar announced that he had succeeded in freezing ordinary air into a clear, transparent solid.[1] This experiment is of considerable importance, inasmuch as neither oxygen nor air had been solidified previous to this time.

The Dewar Vacuum Bulbs

In 1893, Dewar[2] made an important contribution to the researches at low temperatures by the

---

[1] *Chem. News*, 67, p. 126.   [2] *Proc. Roy. Inst.*, 14, p. 1.

introduction of a vessel for storing such volatile
fluids as liquid oxygen, liquid air, etc.   These
liquids, and other similar liquids, which had been
obtained previous to this time, could be kept only
a few minutes in the open
air owing to the rapid
evaporation. Experimen-
ters had already retarded
the evaporation to some
extent by the use of
double-walled   vessels.
Dewar conceived the idea
of exhausting the air from
the space between the
walls of such vessels.

Figure 33 represents
one of these bulbs.   The
outer vessel is exhausted
to a very high degree,
and then sealed off at
the lower end. This pre-
vents, to a considerable

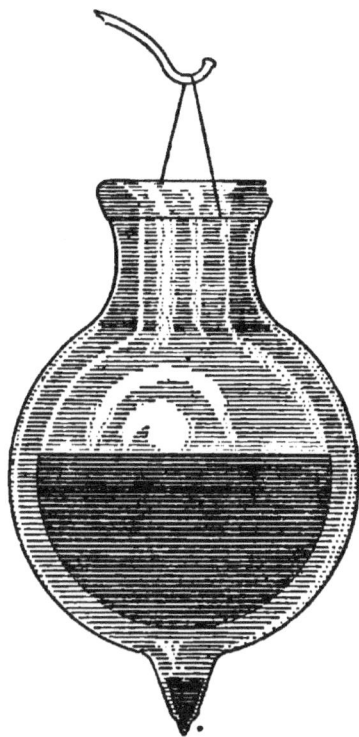

FIG. 33.

extent, the convective transference of heat from
the outer air to the liquid air in the inner vessel.

To test the efficiency of this form of vessel,
comparisons were made with similar vessels which
had not been exhausted. The comparisons were
based upon the amount of liquid oxygen or liquid

ethylene evaporated in a given time. The vessels were placed in water which was kept at constant temperature. The results were as follows : —

	Amount of Gas Evolved
Liquid oxygen surrounded by a vacuum chamber	170 cc. per min.
Liquid oxygen surrounded by an air chamber	840 " "
Liquid ethylene surrounded by a vacuum chamber	56 " "
Liquid ethylene surrounded by an air chamber	250 " "

From these observations it is evident that the rate of evaporation from the air bulbs is about five times that from the vacuum bulbs. By covering the inner vessel with a thin deposit of silver, the rate of evaporation is reduced to less than one half of that given in the table. Liquid oxygen or liquid air can be kept for several hours in such vessels under the ordinary atmospheric pressure.

The next step was to construct a series of vacuum vessels of different forms which could be employed in experiments of different nature. Figure 34 represents a few of the many forms of these vessels which have been constructed. Dewar and other experimenters have made use of these bulbs for investigating the properties of matter at low temperatures, and for determining the properties and physical constants of liquefied gases.

In 1894 Dewar determined the thermal transparency of some liquefied gases for heat of high

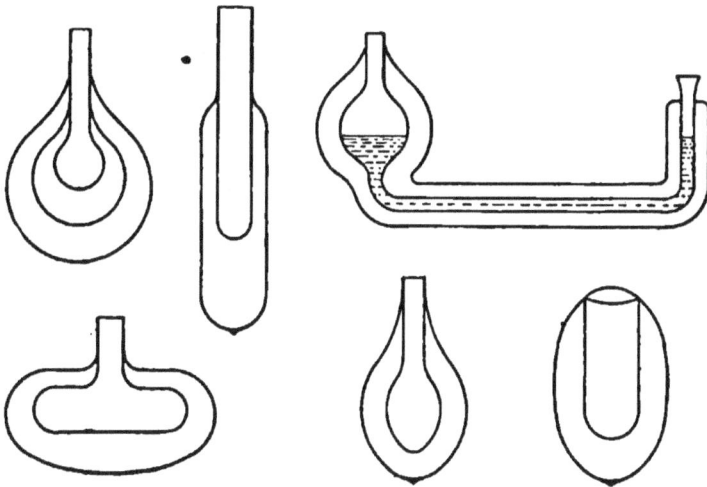

FIG. 34.

refrangibility.[1]   Taking chloroform as the unit of comparison, and correcting for differences in the refractive indices, the results are as follows : —

Chloroform 1.0
Carbon disulphide 1.6
Liquid oxygen 0.9
Liquid nitrous oxide 0 93
Liquid ethylene 0 60
Ether 0.50

[1] *Proc. Roy. Inst.*, 14, p. 393.

Notwithstanding the low temperatures, liquid oxygen and nitrous oxide are very transparent to high temperature radiation.

The author suggested in this article that, instead of silvering the inner bulbs of the vacuum vessels, it is better to leave a small quantity of mercury in the vacuum chambers. When liquid air is introduced into the inner bulbs, a thin coating of solid mercury is found on the outer surface.

A series of observations were also made on the breaking stress of metals at low temperatures. The following table contains the breaking stress, in pounds, for metallic wires of 0.098 in. in diameter : —

	15°	−182°
Steel (soft)	420	700
Iron	320	670
Copper	200	300
Brass	310	440
German silver	470	600
Gold	255	340
Silver	330	420

The breaking stress of wires is not changed by cooling down to −182°, and then allowing the temperature to rise.

In 1895 Dewar made use of a different form of apparatus in the liquefaction of gases.[1] This apparatus involves the regenerative principle in

---

[1] *Proc. Roy. Inst.*, 15, p. 133.

the method of refrigeration, and belongs more properly to the next section (see p. 197).

By means of this apparatus liquid air, etc., can be obtained in a very short time, and at a comparatively small expense. By placing a litre of liquid air in a globular vacuum bulb and subjecting it to exhaustion, the author states that as much as half a litre of solid air can be obtained and maintained in the solid condition for a period of half an hour. An examination of the solid showed it to be a "nitrogen-jelly" containing liquid oxygen.

After obtaining liquid air, oxygen, and nitrogen in considerable quantities, Dewar made an elaborate series of experiments on their densities. The method consisted in weighing different substances of known specific gravities in the liquids. More than twenty substances were weighed in liquid oxygen, and corrections made for the contraction of the solids. The mean of the results gave a value of 1.1375 for the density of liquid oxygen. By weighing a large silver ball in liquid air and liquid nitrogen, the densities of the liquids were found to be 0.910 and 0.850 respectively.

In 1897 Dewar constructed an apparatus for the examination of the least condensible portion of air.[1] From the experiments with this apparatus

[1] *Chem. News*, 76, p. 272.

the author concluded that every gas which occurs
in the atmosphere either condenses to the liquid
state, or is soluble in liquid air.

Numerous observations have been made by
Dewar and Fleming on the electric conductivity
of metals, and the dielectric constant of organic
liquids at low temperatures.[1]

The experiments of Dewar on the liquefaction
of gases since 1895 will be considered in the next
two sections.

EXPERIMENTS OF KAMERLINGH-ONNES

During the last few years Kamerlingh-Onnes
has been experimenting on the liquefaction of
gases. The apparatus which is used by him at
present in the liquefaction of oxygen is somewhat
similar to that employed by Pictet in 1877. The
complete operation consists of three cycles, as
follows:[2] —

First Cycle. Liquid methyl chloride surrounds
a tube containing ethylene under pressure. The
vessel which contains the methyl chloride is connected with an exhaust-pump, by means of which
the pressure is considerably reduced. The rapid
evaporation of the liquid lowers the temperature

---

[1] A list of references to these observations has been compiled by
Dickson. — *Phil. Mag.* [5], 45, p. 528, 1898.

[2] *Zeit. für compr. und flüss. Gase*, 1, p. 169, 1898.

of the chamber, and hence, of the ethylene. The vapors of methyl chloride which pass out of the vessel are compressed, by means of a second pump, in a chamber where they finally condense to a liquid. This liquid passes back again to the vessel which surrounds the ethylene tube, thus making the cycle complete.

Second Cycle. This cycle is similar to the first. The compressed ethylene is liquefied by means of the low temperature which results from the evaporation of the methyl chloride. The liquid ethylene is conducted into a tube, where it surrounds a smaller tube containing compressed oxygen gas. The liquid ethylene is connected with an exhaust-pump, and evaporated under reduced pressure. The resulting vapors are recondensed, and the cycle is made complete.

Third Cycle. In this cycle the oxygen is condensed to the liquid state. The oxygen tube is connected with both an exhaust and a compression pump. The liquid ethylene, boiling under reduced pressure, lowers the temperature of the compressed oxygen to − 140°, when it condenses to a liquid. By means of the pumps the liquid oxygen is removed to a carefully protected reservoir. The operation consists of three complete cycles, and hence is continuous.

A section of the apparatus is shown in the

article to which reference has already been made. The drawing, however, is rather complicated, and for that reason has not been reproduced here. By means of this apparatus the author has obtained liquid oxygen in considerable quantity.

SECTION III

LIQUEFACTION OF GASES BY THE REGENERATIVE METHOD

So far, the low temperatures necessary for the liquefaction of gases have been obtained mainly by the evaporation, especially under reduced pressure, of liquid carbonic acid, liquid nitrous oxide, liquid ethylene, liquid air, etc. A different method of lowering the temperature has become of considerable importance in recent years. The gas which is to be liquefied is compressed to a very high pressure; the heat due to the compression is removed by means of water, and the gas is then allowed to expand. In this way exceedingly low temperatures can be obtained. With this method the gas to be liquefied is the only substance, apart from the apparatus, which is necessary for carrying out the experiment. The process is frequently referred to as the *self-intensification*, or the *regenerative*, method.

The history of this method is rather interesting. According to Joule,[1] Dr. Cullen and Dr. Darwin were the first to observe that the temperature of a gas is lowered by rarefaction and increased by compression. They experimented with ordinary air. Dalton succeeded in measuring this change of temperature with some degree of accuracy (see p. 12). In 1806 Gay Lussac (p. 12) made a series of experiments on the changes of temperature which accompany the compression and rarefaction of gases. Thilorier (1834) observed that carbonic acid could be solidified by means of its own expansion (p. 38). In 1845 Joule, and later Joule and William Thomson, made a very exhaustive study of these phenomena, both experimentally and theoretically. Mayer, Rankine, and Clausius have also contributed largely to this subject. Further reference to these investigations will be made in the theoretical considerations which will be taken up later.

In connection with these and many other similar observations which were conducted solely for scientific purposes, a number of experiments have also been made with a view of applying this method of lowering the temperature for refrigerating purposes. In 1849 Dr. John Gorrie[2] constructed an

---

[1] *Phil. Mag.* [3], 26, p. 369.
[2] Wallis-Tayler, *Refrigerating and Ice-Making Machinery*, p. 116.

ice machine, based upon the compression and expansion of air. In 1857 Siemens[1] constructed a refrigerating machine, which he describes as follows : —

" The invention relates to freezing and refrigerating by the expansion of air or elastic fluid. The air is first compressed by a cylinder, or by pumps of any suitable construction, by which the temperature is raised, and it is cooled while in the compressed state, and is then allowed to expand in a cylinder or engine of any suitable construction, by which the temperature is lowered. The air thus cooled is brought in contact with the articles to be cooled or frozen, and is then conducted through an interchanger, or apparatus, by which it is made to cool the compressed air which enters the interchanger in the opposite direction. . . . The principle of the invention is adapted to produce an accumulated effect, or an indefinite reduction of temperature." Kirk in 1863, Marchant in 1869, and Giffard and Postle in 1873, constructed ice machines in which the freezing was produced by the expansion of air.[2] In 1885 Solvay[3] described a method which is similar to that of Siemens. A

---

[1] Linde, *Engineer*, 82, p. 486.

[2] Wallis-Tayler, *Refrigerating and Ice-Making Machinery*, p. 116 *et seq.*

[3] Linde, *Engineer*, 82, p. 486.

very efficient refrigerating machine, based upon this same principle, was constructed by Windhausen.[1] Reference might be made to other machines of a similar nature, but it is not within the scope of a book of this kind to consider the subject of refrigerating machines. Those who are interested in that branch of the work can refer, in connection with the various text-books and journals on engineering, to Wallis-Tayler's *Refrigerating and Ice-Making Machinery*, and especially to the two journals *Ice and Refrigeration*, published in New York and Chicago, and the *Zeitschrift für die gesammte Kälte-Industrie*, published in Leipzig.

The preceding examples, however, are sufficient to show that the lowering of temperature, which accompanies the expansion of a gas, was observed a century ago, and that this method of lowering the temperature has been applied technically for at least fifty years. The method is not new.

In 1874 Edwin J. Houston[2] suggested a form of apparatus which is very similar to the machines employed at present in the liquefaction of air. The following are his words: "The means of obtaining exceedingly low temperatures seem at last to have been fulfilled in the 'Windhausen Ice and Refrigerating Machine.' Though introduced

---

[1] Houston, *Jour. Frank. Inst.*, 67, p. 10.
[2] *Ibid.*, p. 9, 1874.

for practical purposes, mainly for the cheap production of artificial ice, the machine contains latent
possibilities, which we hope will at once be utilized,
that open up the most promising field to the original investigator, and bid fair to enrich science with
stores of new facts."

" . . . In the Windhausen process, a steam
engine is employed to compress the air to 2 or 3
atmospheres. The heat developed by the compression is drawn off during the passage of the
condensed air through pipes in a series of chambers, in which cold water is flowing. The cooled
air is then allowed to expand into a cylinder under gradually diminishing pressure, the expansion
being attended with the development of great
cold."

" . . . The following modifications of the apparatus would render its cold-producing power
almost unlimited : —

" 1. A communication between the expansion
cylinder and the chambers through which the condensed air is conducted before it is allowed to
expand. Supposing this outlet regulated by a
cock, a blast of very cold air could replace the run-
ning water, and reduce the compressed air to a
very low temperature.

" 2. The introduction of a second compressing
cylinder, with which the compressed air, after be-

ing cooled, could be still further compressed, again cooled, and finally conducted into the expansion cylinder. Under a pressure of, say, 60 atmospheres, a considerable mass of air at the temperature of, say, $-100°$ F. would, in its expansion, produce a reduction of temperature greater perhaps than any yet obtained. . . . There would appear to be no other limit to the reduction of temperature save what would arise from the strength of materials, or the liquefaction and subsequent freezing of the nitrogen, or the oxygen of the air, or of the air itself.

" Among the advantages that we may rationally expect to accrue from the apparatus thus modified are the following : —

" 1. The confirmation or otherwise of the ' absolute zero ' as determined by the expansion or contraction of gases, by heat or cold.

" 2. The liquefaction and subsequent solidification of many of the incoercible gases, the determination of their physical peculiarities as liquids or solids, together with their crystalline form.

" 3. The action of intense cold on the chemical affinities of certain gaseous compounds.

" 4. The action of intense cold on the color of certain chemical compounds."

These predictions have been thoroughly fulfilled. In the apparatus suggested, Houston anticipated

the methods which were employed twenty years
later in the liquefaction of air by Linde, Hampson,
and Tripler.

In 1875 Coleman[1] constructed an apparatus for
the liquefaction, on a large scale, of some very
volatile hydrocarbons. The lowering of the temperature was produced by the expansion of the
compressed gases. The machine involves : —

" 1. The pumping of the gas by steam power
into a system of tubes capable of being externally
cooled, and from which condensed liquids can be
drawn off by ball-cocks.

" 2. Employing the compressed gas, after being
deprived of its liquid, for working a second engine
coupled with, and parallel to, the first, thus receiv-
ing a portion of the force originally employed in
the compression.

" 3. Employing the expanded gas, after having
had its temperature reduced in the act of doing
the work of pumping, for supplying the necessary
cold for cooling a portion of the condenser pipes
to zero."

The cycle of operations was complete. The
temperature of the compressed gas was about 5°,
while the minimum temperature after expansion
was −45°. With this apparatus a continuous

---

[1] *Chem. News*, 39, p. 87.

stream of the liquefied gas could be produced. About 250,000 gallons of the liquid hydrocarbons were produced during the first three years.

Cailletet and Pictet have also made use of the expansion of gases as a means of lowering the temperature, but in their work the regenerative principle was not used. Wroblewski, Olszewski, and Dewar, previous to 1895, made experiments of a similar nature. Reference has already been made to these observations. In 1894 Kamerlingh-Onnes made use of the regenerative principle in the liquefaction of gases.[1] Omitting a number of observations on the compression and expansion of gases, which have but slight bearing on the condensation of gases, we may proceed at once to the consideration of some forms of apparatus which have recently been constructed for the purpose of liquefying oxygen, nitrogen, and especially ordinary air.

APPARATUS EMPLOYED BY LINDE IN THE
LIQUEFACTION OF AIR

This apparatus is an outgrowth of a long experience in the construction of refrigerating machines. We may omit these devices, however, and proceed at once to the consideration of the apparatus employed in the liquefaction of air.

[1] Ref. Dewar, *Proc. Roy. Inst.*, 15, p. 133.

The apparatus was successfully operated in May, 1895, and was constructed as follows :[1] The general arrangement of the apparatus is shown in figure 35. The air to be liquefied is brought by means of the compressor C to a pressure of about 200 atmospheres (in the first few experiments the pressure employed was about 65 atmospheres). R is a water-cooler, which serves to remove the heat of compression. The compressed air then passes through the concentric tube-system H, which is placed in a large, well-insulated chamber, to the expansion valve r, where it is allowed to escape into the receiver G. The sudden expansion of the air produces a considerable lowering of temperature. The cold air rushes out of the receiver, and passes up through the outer tube, thus lowering the temperature of the compressed air in the inner tube. As the process continues the temperature gradually falls until the air begins to liquefy in the receiver G. This takes place at the temperature of boiling air ; consequently, the liquid air is obtained in a static condition at the ordinary atmospheric pressure.

In this way Linde obtained liquid air in considerable quantity. The resulting liquid was very rich in oxygen, inasmuch as the boiling point of oxygen is

higher than that of nitrogen.   The author modified
the apparatus somewhat in order to obtain a more
complete separation of the oxygen and nitrogen.[1]
The   apparatus   employed   at   present [2] differs

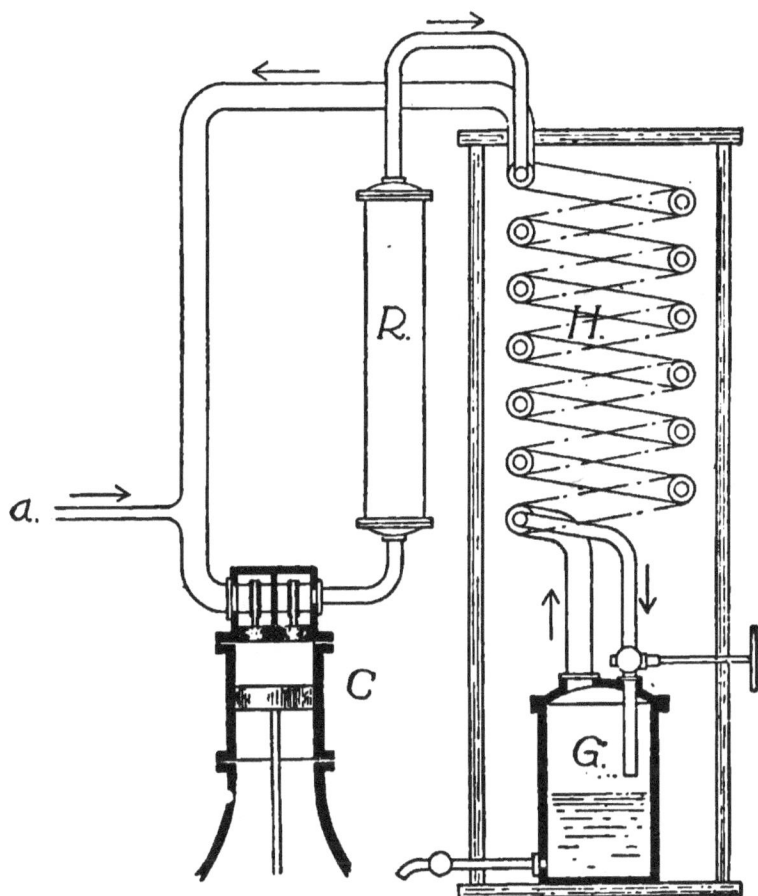

FIG. 35.

[1] *Engineer*, 82, p. 509.   The efficiency and theory of the apparatus are also considered by Linde in this article.

[2] *Zeit. f. Elektrochem.*, 4, p. 4, 1897 ; also *Zeit. Compr. und
flüss. Gase*, 1, p. 117, 1897.

somewhat from that just described. The general
arrangement of this apparatus is shown in figure
36. The tube-system in this case consists of three

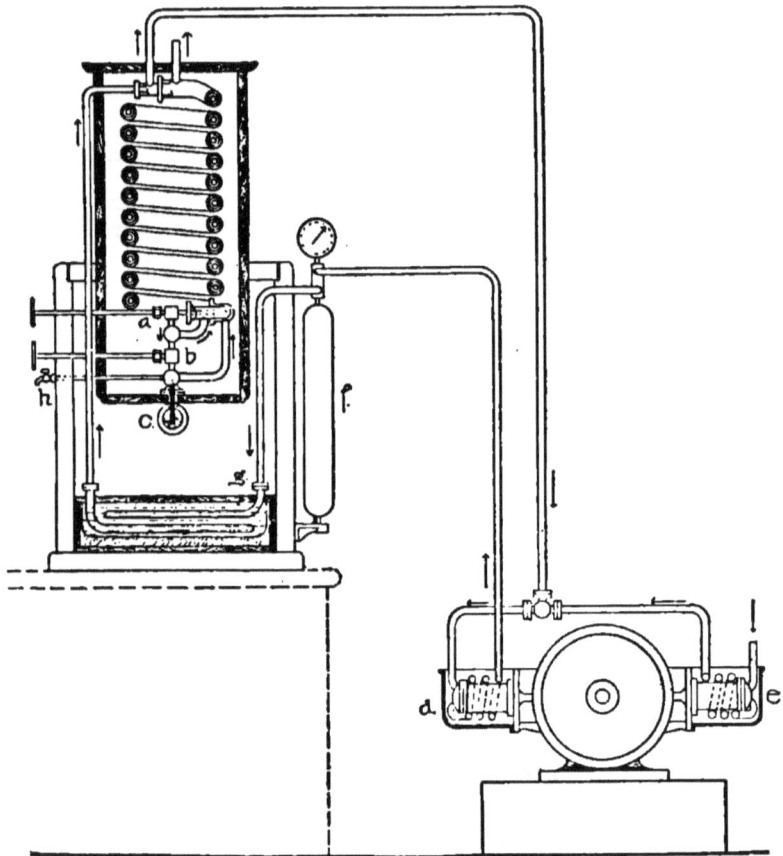

FIG. 36.

concentric copper tubes, arranged in the form of
a spiral. The coils are placed in a carefully in-
sulated chamber. By means of the compressor
d, the air is forced into the innermost tube of the

system under a pressure of about 200 atmospheres. The heat of compression is removed by
means of the water-cooler G. The small tube is
provided with an expansion valve at a, where the
air is allowed to expand into the space between
innermost and middle tubes. In this operation
the air passes from a pressure of 200 atmospheres
to a pressure of 16 atmospheres. The temperature, of course, is greatly reduced by the sudden
expansion of the air. The cold air passes up
through the middle tube, and finally back to the
compressor d, where it is again compressed to
200 atmospheres. This process can be seen from
the directions of the arrows. The compressor e
maintains a pressure of 16 atmospheres in the
middle tube. As the process continues, the temperature gradually becomes lower. The middle
tube is provided with an expansion valve at b,
similar to that at a, through which the air is allowed to expand a second time. In this case, the
air passes from a pressure of 16 atmospheres into
the space between the outer and middle tubes,
where the pressure is one atmosphere. This
operation lowers the temperature still further.
The cold air passes up through the outer tube of
the system, abstracting heat from the compressed
air, and finally escapes through the outlet at the
top of the chamber. The apparatus is so adjusted

that the compressor *e* supplies sufficient air to
exactly replace the small quantity which is lost.
The cycle is now complete, and the temperature
of the system gradually decreases. When the process has continued for a period of from $1\frac{1}{2}$ to 2
hours, liquid air begins to collect in the bottom
of the chamber, and passes into the double-walled
Dewar bulb *c*. The liquid can be drawn from this
vessel by opening the stop-cock *h*. About one
litre of liquid air can be produced per hour, with
a three horse-power engine. With a larger apparatus, and a correspondingly increased power,
the liquid, of course, could be obtained in much
larger quantities. Numerous observations have
also been made by this experimenter on the properties of matter at low temperatures.

Linde has in the process of construction, for
the Rhenania Chemical works at Aix-la-Chapelle,
a liquid air plant with a capacity of about 50 litres
of liquid air per hour, and an expenditure of about
120 horse-power.[1]

HAMPSON'S APPARATUS FOR THE LIQUEFACTION OF GASES

The apparatus employed by Hampson in the
liquefaction of oxygen, air, etc., is based upon the

---

[1] Ewing, *Engineering*, 65, p. 310, 1898.

same principle as
that employed . by
Linde, which has just
been described. Con-
siderable discussion
has arisen concern-
ing this form of ap-
paratus as to priority
of invention. Hamp-
son's patent is dated
May 23, 1895. With
this statement as to
the date of the inven-
tion, we may leave
the question of pri-
ority and proceed at
once with a descrip-
tion of the apparatus.

A longitudinal sec-
tion of the apparatus[1]
is represented in
figure 37. The gas
to be liquefied (oxy-
gen, air, etc.) is in-
troduced through the
small tube at the

[1] *Engineer*, 81, p. 310,
1896.

o

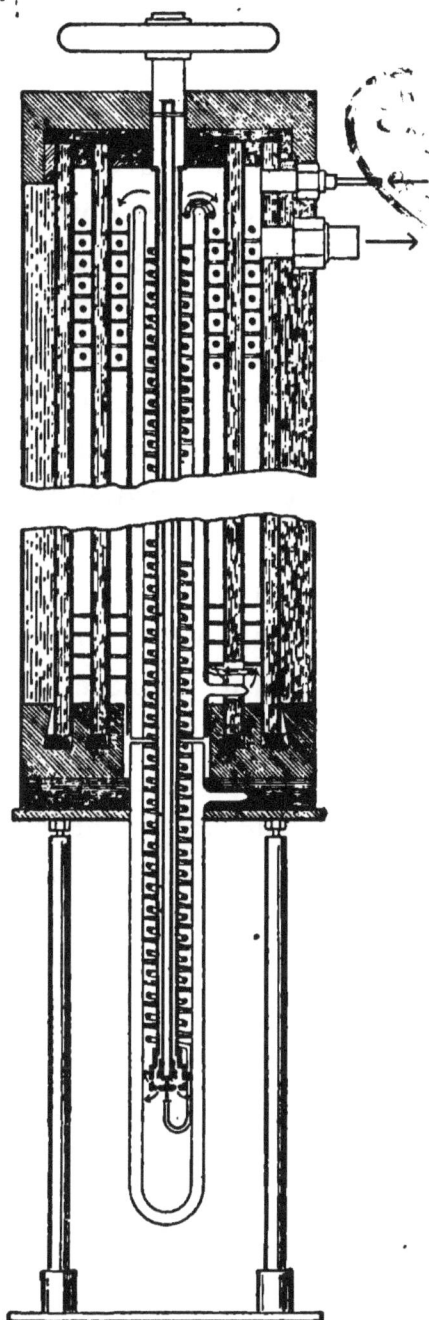

FIG. 37.

upper end under a pressure of 120 atmospheres. The gas enters the apparatus at the ordinary temperature (the heat of compression having been removed). The tube through which the gas is introduced extends downward in the form of a spiral around a central column. At the lower end of the spiral is an expansion valve which opens from above by means of a screw. The expanded gas passes up through the apparatus and out again at the upper end. The outer portions of the apparatus are carefully packed with felt.

The details of the spiral tube and expansion valve are shown in figure 38. The coil *A* terminates in the jet-piece *D*, which delivers the gas against a flat plug on the screw *C*. A slightly different form of valve is shown in the small figure to the left. The gas on escaping through this valve expands to a pressure of one atmosphere. The temperature of the gas is considerably reduced in consequence of this expansion. The cold gas passes up around the coil as shown by the arrows in the figure, thus lowering the temperature of the compressed gas. This makes the process self-intensifying. As the operation continues the temperature gradually falls, until finally the gas begins to liquefy. Liquid oxygen and liquid air can be obtained in this way in a

static condition at the ordinary atmospheric pressure in a comparatively short time.   The apparatus
was constructed on
rather a small scale, and
had a capacity of about
two cubic centimetres
of liquid oxygen per
minute.

During the next year
the apparatus was some-
what modified.[1] The air
in this case was intro-
duced under a pressure
of 87 atmospheres. The
jet of liquid air could
be seen within twenty-
five minutes, and the
liquid air began to col-
lect in the receiver
within thirty-three min-
utes from the beginning
of the operation.   A
vacuum vessel was used
as the receiver.

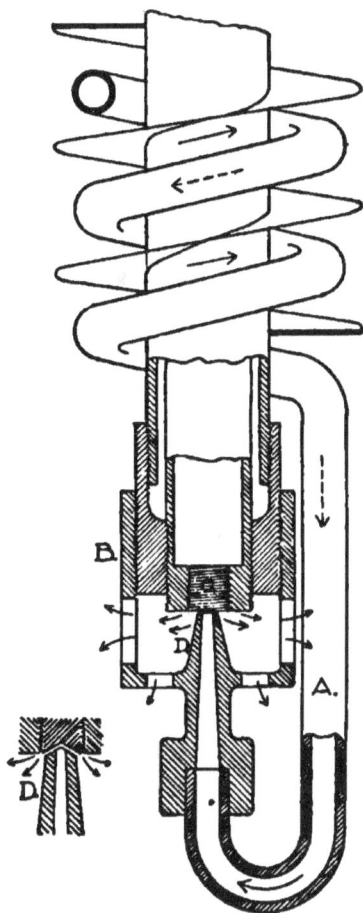

FIG. 38.

[1] *Engineer*, 83, p. 294.

APPARATUS EMPLOYED BY DEWAR IN THE
LIQUEFACTION OF GASES

On the 19th of December, 1895, Dewar read a
paper before the English Chemical Society on the
"Liquefaction of Air and Researches at Low
Temperature."[1] In this paper the author describes an apparatus in which the self-intensification method of refrigeration is employed. · The
compressed gas is cooled to about − 80° by means
of liquid carbonic acid, and then allowed to expand under a regenerative coil. A section of the
apparatus is shown in figure 39.

The carbonic acid is introduced into the apparatus at *B*, and passes through the coiled tube represented by the shaded circles. *C* is a carbonic
acid valve, and *H* the carbonic acid outlet. The
air or oxygen to be liquefied enters the apparatus
at *A*, under a pressure of 150 atmospheres. *D* is
a regenerative coil, and *F* an expansion valve
where the compressed gas is allowed to expand
into the vacuum vessel *G*.

When the temperature of the system is lowered to
about − 80°, the gas is allowed to expand through
the valve *F*. This produces a further decrease
in the temperature. The cooled gas passes up
around the regenerative coil and lowers the tem-

[1] *Proc. Roy. Inst.*, 15, p. 133.

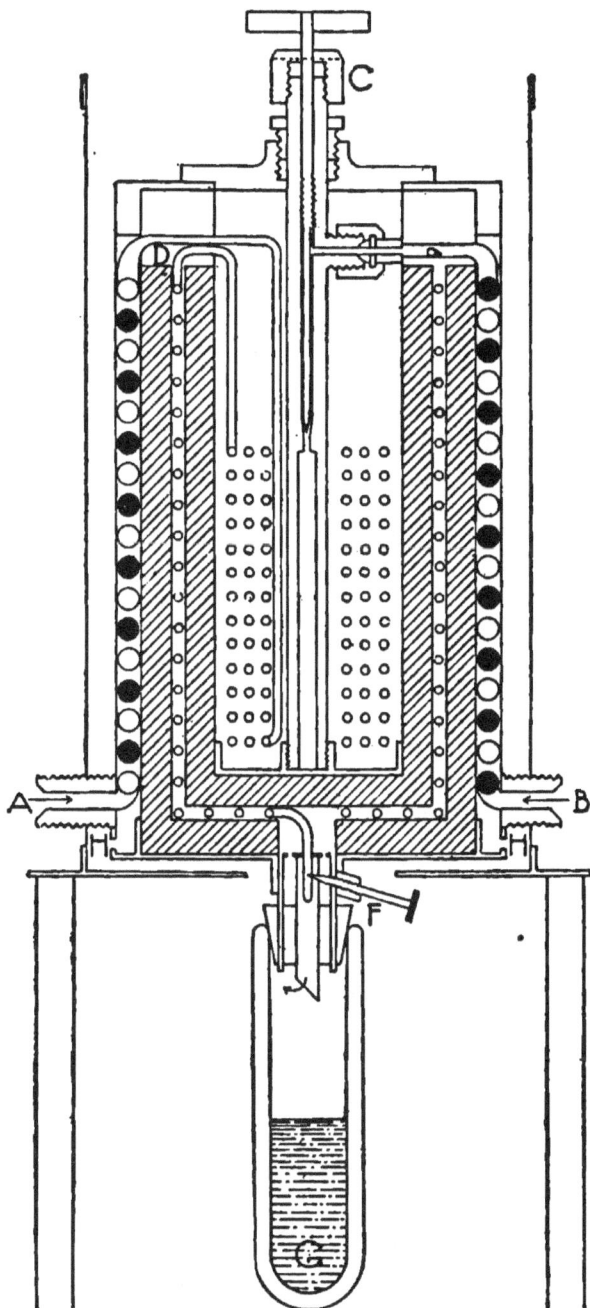

FIG. 39.

perature of the compressed gas. The temperature
of the system falls rapidly, and within fifteen
minutes from the time of starting, liquid air (or
whatever the substance may be) begins to drop
into the vacuum vessel G. By means of this
apparatus Dewar was able to liquefy oxygen, air,
etc., on a comparatively large scale.

Dewar also made a number of experiments on
"gas jets containing liquid." In these observations a small regenerative coil was placed in a long
vacuum vessel. Three different forms of these
vessels were constructed. The compressed gas
was allowed to expand from the coil near the
lower end of the vacuum tube. Oxygen under a
pressure of 100 atmospheres, and previously cooled
to $-79°$, produced a visible jet of liquid. Dewar
suggests this method as a very rapid means of
obtaining low temperatures.

Similar experiments were made with hydrogen.
When cooled to $-200°$ and allowed to expand from
a pressure of 140 atmospheres in one of these
tubes, no liquid jet could be seen. If the gas contained a few per cent of oxygen the liquid jet became visible. By allowing the pure hydrogen to
expand from the pressure of 200 atmospheres over
a longer regenerative coil, which has been previously
cooled to $-200°$, the liquid jet becomes visible.
The liquid hydrogen, however, could not be ob-

tained in a static condition. To obtain some idea of the temperature of the liquid jet, the author placed some liquid air and liquid oxygen in the lower end of the tube. Within a few minutes about fifty cubic centimetres of these liquids were transformed into hard, white solids, resembling avalanche snow. The solid oxygen had a pale bluish color, and showed by reflection all the absorption bands of the liquid.

Tripler's Apparatus for the Liquefaction of Air

The method employed by Tripler in the lique- faction of air is based upon the same principle as that employed by Linde and Hampson, but is operated on a much larger scale. Unfortunately no thoroughly scientific account of this apparatus has yet been published. The temperature is low- ered by expanding highly compressed air in a long tube under a regenerative coil. The general plan of the apparatus is shown in figure 40.[1]

The steam boiler *b* supplies steam to the com- pressor *c* under a pressure of about 85 pounds per square inch. The compressor is provided with three air cylinders arranged in tandem on the same rod. These cylinders are cooled by means

[1] *Engineering News*, 39, p. 246, 1898.

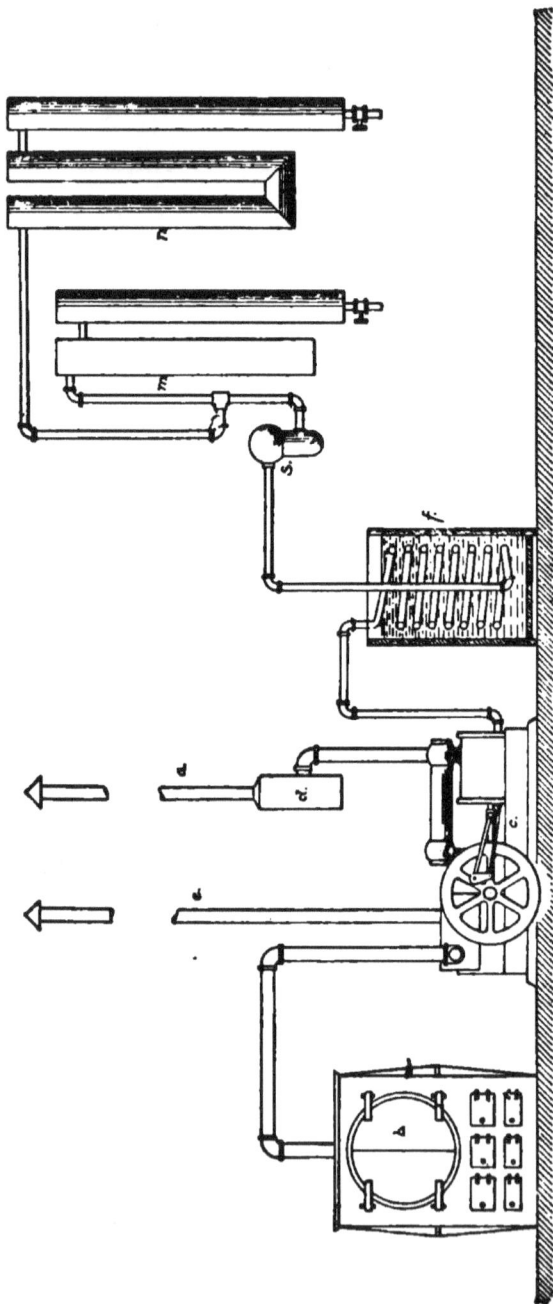

FIG. 40.

of water-jackets. The first or low-pressure cylinder is 10½ inches in diameter, and compresses the
air to a pressure of about 4 atmospheres. The
second or intermediate cylinder is 6⅞ inches in
diameter, and increases the pressure to about 25
atmospheres. The third or high-pressure cylinder
is 2⅝ inches in diameter, and delivers the air at
a pressure of about 150 atmospheres. The air
is taken into the compressor from the outside
through the pipe *a*, which is provided with a dust-
separator at *d*. The heat of compression is removed by means of the water in the tank *f*. After leaving the cooler the compressed air passes
through the separator *s*, which removes the last
traces of moisture.

The air is now ready to be liquefied. This is
accomplished by means of the two liquefiers *m*
and *n*. These liquefiers consist, in each case, of a
long coil of copper tubing with an expansion valve
at the lower end. The coils are carefully protected from the external heat by means of felt.
The air enters these liquefiers under a pressure
of 150 atmospheres, and is allowed to expand to a
pressure of one atmosphere. The temperature is
considerably reduced by this expansion; the cold
air passes up around the coil and lowers the temperature of the compressed air, and thus the influence becomes accumulative. Within about twenty

minutes the air begins to liquefy in the lower
end of the tube. The liquid air is removed by
means of the valves at the lower ends of the
liquefiers.

As operated at present this apparatus has a
capacity of from three to four gallons of liquid air
per hour. Tripler has obtained liquid air on a
much larger scale than have any of the other
experimenters. It is not an uncommon occurrence to ship liquid air from this plant (in ten-
gallon cans) to a distance of several hundred
miles. The liquid has been successfully trans-
ported from New York to Philadelphia, Wash-
ington, Boston, and other cities. This plant has
made it possible to experiment with liquid air in
both technical and scientific lines at a comparatively small expense.

THEORY OF THE SELF-INTENSIFICATION METHOD OF REFRIGERATION

An elaborate discussion of this problem necessi-
tates the use of complicated mathematical formulæ,
and would be out of place in a work of this nature.
For that reason the subject will be considered
only in a general way.[1]   It has been known for

---

[1] For a more complete discussion, see Joule and Thomson,
Trans. Roy. Soc., 144, p. 321, or some work on thermodynamics.

more than a century that when a gas is compressed, the temperature rises, and, conversely, when the compressed gas is allowed to expand, the temperature falls. In 1842 Mayer investigated the cause of these thermal changes. "Whence comes the heat," he asks, "generated during the compression of a gas, and what becomes of the heat which vanishes when the gas expands?" He considered the problem in the light of the conservation of energy. The work which is done in the compression of a gas, he says, is changed into heat, and the work which is done by the expanding gas, against the external pressure, cannot spring into existence from nothing, but comes from the heat which the gas loses. Mayer's conception implies that heat is a form of energy. He fully recognized the fact that heat can be obtained from or transformed into other forms of energy.

In order to make this subject clear, it will be necessary to consider the two specific heats of gases. By the specific heat of a substance is meant the capacity of unit mass of the substance for heat; *i.e.* the ratio of the amount of heat supplied to the body to the rise in temperature. When a gas is heated, either the pressure or volume is increased. If the pressure of a gas is kept constant during a specific heat determination, the value obtained is greater than that obtained at constant volume. In the former case the gas expands against a definite pressure; *i.e.* it performs a certain amount of mechanical work. The excess of heat required to raise the temperature of a gas at constant pressure over that required at constant volume is simply the amount of energy required in the expansion of the gas.

According to this theory there should be no temperature change when a gas is allowed to expand into a vacuum, as there is no external work to overcome. The experiments of Gay Lussac and those of later experimenters have shown this to be true.[1] Moreover, if the theory of Mayer is true, the difference between the quantities of heat necessary to raise the temperature of one gram of air through one degree under the two conditions; *i.e.* at constant pressure and constant volume, should correspond to the work performed in expanding one gram of air $\frac{1}{273}$ of its volume at 0°. Both of these values have been determined. The first is 0.0692 calories of heat, and the second 2923.5 gram centimetres of work. If, then, 0.0692 calories of heat correspond to 2923.5 gram centi-

---

[1] Strictly speaking, this statement is true only in the case of a perfect gas. Although there is no external work to overcome when ordinary gases expand into a vacuum, a small quantity of energy is required to overcome the molecular attraction. This causes a slight decrease in the temperature.

metres of work, one calorie of heat corresponds to $\frac{2923.5}{0.0692} = 42245$ gram centimetres of work. This latter value, being equivalent to one calorie, is called the mechanical or dynamical equivalent of heat, and agrees very closely with the values obtained by different methods by Joule, Rowlands, and others. We may safely assume, then, that the heat which is absorbed when a compressed gas is allowed to expand is equivalent to the external work performed.

In order to calculate the decrease in temperature which accompanies the expansion of a gas, it is necessary to know the specific heat of the gas at constant pressure and constant volume; also the initial and final pressures of the gas.

Suppose a small quantity of heat, dQ, be communicated to a gas at constant volume, and let the rise in temperature be represented by dT. Then, as no external work has been done,

$$\frac{dQ}{dT} = C_v, \qquad (1)$$

where C_v represents the specific heat of the gas at constant volume. Equation (1) may be written

$$dQ = C_v\,dT.$$

If the pressure p had remained constant when the heat was communicated, there would have been a slight increase in volume which we may represent by dv. Then a certain amount of external work (pdv) would have been performed. If this vaiue. is represented in thermal units, the available heat for increasing the temperature of the gas is $dQ - pdv$, and the above equation becomes

$$dQ - pdv = C_v\, dT \qquad (2)$$

or
$$\frac{dQ}{dT} = C_v + \frac{pdv}{dT}.$$

The term $\dfrac{dQ}{dT}$ in this case represents the specific heat of the gas at constant pressure, and hence

$$C_p = C_v + \frac{pdv}{dT}, \qquad (3)$$

where C_p is the specific heat at constant pressure.

By differentiating the equation $pv = RT$ with respect to v, we obtain

$$pdv = RdT,$$

or
$$R = \frac{pdv}{dT};$$

hence
$$C_p = C_v + R.$$

If the equation $pv = RT$ be differentiated with respect to all of the variables, we have

$$pdv + vdp = RdT.$$

Substituting for R its equivalent $C_p - C_v$, this equation becomes

$$dT = \frac{pdv + vdp}{C_p - C_v}.$$

By substituting this value of dT in 2, we obtain

$$dQ = \frac{C_p}{R} pdv + \frac{C_v}{R} vdp. \qquad (4)$$

If the gas is allowed to expand adiabatically,[1] $dQ = 0$, and hence, from 4, we have

$$\frac{C_p}{R} pdv + \frac{C_v}{R} vdp = 0.$$

Dividing this equation by $\frac{C_v}{R}$, we obtain

$$\frac{C_p}{C_v} pdv + vdp = 0.$$

The last equation may be written

$$k \frac{dv}{v} + \frac{dp}{p} = 0, \qquad (5)$$

where k represents the ratio of the specific heat of the gas at constant pressure to that at constant volume. If the gas be allowed to assume two definite conditions with respect to pressure and volume, p, v, and p_1, v_1, and we integrate equa-

---

[1] A gas is said to expand adiabatically when no heat is communicated from or given out to the external surroundings.

tion (5) between these limits, and transpose the result, we obtain

$$\log p - \log p_1 = k\,(\log v_1 - \log v),$$

or
$$k = \frac{\log\left(\dfrac{p}{p_1}\right)}{\log\left(\dfrac{v_1}{v}\right)}$$

Simplifying this, we have

$$\frac{v_1}{v} = \left(\frac{p}{p_1}\right)^{\frac{1}{k}}. \tag{6}$$

Suppose now that a gas be allowed to expand from the pressure p to the pressure p_1. Before and after the expansion the gas must fulfill the equations $pv = RT$, and $p_1 v_1 = RT_1$, respectively. Dividing the first of these by the second, we have

$$\frac{pv}{p_1 v_1} = \frac{T}{T_1}.$$

If we substitute for $\dfrac{v}{v_1}$ its value from equation (6), the equation becomes

$$\left(\frac{p}{p_1}\right)^{\frac{k-1}{k}} = \frac{T}{T_1}. \tag{7}$$

Knowing, then, the initial and final pressures, and also the initial temperature, the final temperature can be calculated from equation (7).

In the case of air, $k = 1.41$. Substituting this

value, and assuming the initial temperature to be zero centigrade and the final pressure to be one atmosphere, the theoretical temperatures obtained by the adiabatic expansion of air are given in the following table : — [1]

Initial Pressures	Final Temperatures
100 atmospheres	− 201°.5
200 "	− 214 .5
300 "	− 221 .0
400 "	− 225 .1
500 "	− 228 .2

Applying these results to the liquefaction of gases by means of the regenerative coil, it is evident that the expansion of the gas in the tube lowers the temperature by an amount which corresponds to equation (7). Further, the issuing jet experiences a much greater decrease in temperature owing to the greater difference between the initial and final pressures. Finally, the expanded gas of very low temperature passes up around the coil and lowers the temperature of the compressed gas ; *i.e.* the initial temperature of the gas, before expansion, gradually decreases. In this way the influence becomes accumulative, and the

[1] Ostwald's *Outlines of General Chemistry*, p. 83.

P

process has been termed the regenerative or self-intensifying method.

In practice the efficiency of such an apparatus is never equal to the theoretical. In the first place, there is always an inflow of heat from the outside. Careful insulation may reduce, but can never entirely eliminate, this influence. Furthermore, the interchange of heat between the expanded gas and the counter-current apparatus is never complete.[1]

Raylcigh suggests that the efficiency of the various forms of apparatus employed at present in the liquefaction of air might be considerably increased by the use of a turbine.[2] He says: "It must not be overlooked, that to allow the work of expansion to appear as heat at the very place where the utmost cooling is desired, is very bad thermodynamics. The work of expansion should not be dissipated within, but be conducted to the exterior. . . . A turbine of some sort might be used. This would occupy little space, and even if of low efficiency, would still allow a considerable fraction of the work of expansion to be conveyed away. The worst turbine would be better than none, and would probably allow the pressures to

[1] For details in regard to the efficiency of the apparatus, see Linde, *Engineer*, 82, pp. 485 and 509.

[2] *Nature*, 58, p. 199, 1898.

be reduced. It should be understood that the object is not so much to save the work, as to obviate the very prejudicial heating arising from its dissipation in the coldest part of the apparatus. It seems to me that the future may bring great developments in this direction, and that it may thus be possible to liquefy even hydrogen at one operation."

Since the regenerative principle has proved successful in the liquefaction of air, it has been suggested by some inventors that the immense quantity of heat which is stored up in the earth's atmosphere may be obtained as available energy for doing work at a very small expense. Their idea is that there is considerably more energy stored up in the liquid air than is required to produce the liquid by the regenerative method. Without going into the theory of this suggestion, we may dispose of it by quoting the following passage from Nernst:—

"In the judgment of some inventors who are completely permeated with the accuracy of the law of the conservation of energy, it is by no means regarded as impossible to construct a machine which should be able to furnish work as desired and free of cost. External work and heat are equivalent to each other. Moreover, energy in the form of heat is in abundance, so that it only needs an apparatus in which one shall apply it in driving our machine, to use up the energy of its environments. Such an apparatus, for example, might be sunk into a great water reservoir, whose enormous quantity of energy could be changed into useful work; it would, for example, make the steam engines of our ocean steamers unnecessary, and would keep the screw of a ship in motion as long as desired, and at the cost of the immeasurable store of heat in the sea. Such an apparatus would be in certain respects a perpetual motion, and yet not contradictory to the first law of thermodynamics, since it would extract the heat of its environments and give it back again as external work which, as a result of the friction of the screw, would change itself back again into heat, to enter the cycle anew.

" Unfortunately such an apparatus, which would make coal worthless as a source of energy, appears to be a chimera, exactly as was the perpetual motion of the inventor of the last century; at least, many fruitless attempts have made this more than probable. Thus, as we sum up the numerous abortive endeavors, we come, in a way analogous to that which led .to the knowledge of the conservation of energy, to the proposition that an apparatus which could continually change the heat of its environments into external work is a contradiction to a law of nature, and therefore an impossibility. Although by recognizing this law the human spirit of invention may be poorer by one problem, yet natural investigation is compensated for it by a principle of almost unlimited application."

SECTION IV

LIQUEFACTION OF ARGON, HYDROGEN, HELIUM, ETC.

There remain to be considered some special observations in the liquefaction of gases. The quantity of gas to be liquified, in some cases, was very small, and hence an apparatus of special construction was necessary.

Liquefaction and Solidification of Argon

In 1895 Olszewski[1] subjected argon to low temperatures and high pressures. A sample of the pure dry gas, about 300 cc., was obtained from Ramsay. Four series of experiments were made; two with the object of determining the critical temperature and pressure, and two for the purpose of determining the boiling point and freezing point under atmospheric pressure.

---

[1] *Trans. Roy. Soc.*, 186, p. 253, 1895.

In the first two experiments use was made of an apparatus similar to that of Cailletet. Liquid ethylene, boiling under reduced pressure, was used as the refrigerant. At a temperature of $-128°.6$ and a pressure of 38 atmospheres the argon condensed to a colorless liquid. On slowly raising the temperature the meniscus became less distinct and finally disappeared. This was repeated several times with a view of determining the critical temperature and pressure. The mean of seven observations gave $-121°$ for the critical temperature, and 50.6 atmospheres for the critical pressure. The vapor pressure of argon at a temperature of $-139°.1$ was found to be 23.7 atmospheres.

The experiments on the boiling and freezing points were carried out by means of the apparatus represented in figure 41. The argon was contained in the glass burette b, closed at both ends with glass stop-cocks. The lower end of the burette was connected, by means of a flexible tube, with the mercury reservoir a. By means of the mercury, the gas could be transferred to the liquefying tube d, which was immersed in the liquid oxygen contained in the quadruple-walled glass tube e. The tube i was connected with a large air-pump, and the tube c with a mercury air-pump.

When the temperature of the tube d had become equal to that of liquid oxygen boiling under the ordinary atmospheric pressure, the argon was

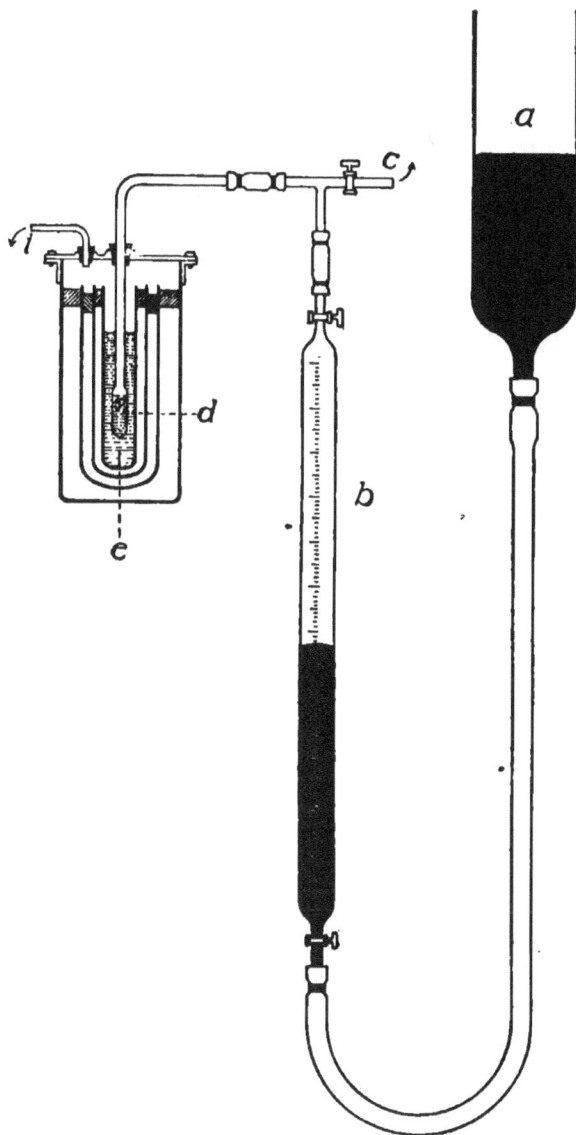

FIG. 41.

admitted, but showed no signs of liquefaction. This showed that the boiling point of argon is lower than that of oxygen. The pressure of the argon was then adjusted so as to remain equal to that of the atmosphere. The tube containing the liquid oxygen was slightly exhausted, and at a temperature of − 187° the argon began to liquefy. As a mean of four experiments Olszewski gives − 187° for the boiling point of argon. From the volume of gas used and the volume of liquid obtained, the density of liquid argon at the boiling point was estimated to be about 1.5.

The temperature of the liquid oxygen was then lowered by slow exhaustion to − 191°, when the argon solidified to a crystalline mass resembling ice. At lower temperatures the mass became opaque. The substance was frozen and melted four times. The mean value obtained for the melting point by means of a hydrogen thermometer is − 189°.6.

Experiments with Helium

In 1896 Olszewski[1] made an extensive series of experiments with a view of liquefying helium. The pressure was obtained by means of a Cailletet apparatus. The liquefying tube of this appa-

---

[1] *Wied. Ann.*, 59, p. 184 ; *Nature*, 54, p. 377.

ratus was thoroughly exhausted by means of a mercury pump, and then carefully filled with dry helium which had been obtained from Ramsay. In the first series of experiments liquid oxygen was used as the refrigerant. By the evaporation of this liquid under a pressure of 10 mm. of mercury a temperature of − 210° could be obtained. At this temperature and under a pressure of 125 atmospheres helium showed no signs of liquefaction. The gas was then allowed to expand until the pressure had decreased to twenty atmospheres, and in some cases to one atmosphere, but there was no evidence of condensation to the liquid state.

In the second series of experiments liquid air was employed as the refrigerant. By the evaporation of liquid air under a pressure of 10 mm. of mercury the temperature was reduced to about − 220°. Even at this low temperature and with a pressure of 140 atmospheres, the results were all negative. Under these conditions the gas was again allowed to expand, but there were no indications of liquefaction. The author says : " In every single instance I have obtained negative results, and, as far as my experiments go, helium remains a permanent gas, and apparently much more difficult to liquefy than even hydrogen." The experiments were carried out on a very small scale, owing to the small quantity of gas at hand (about 140 cc.).

The temperatures were not measured directly in these experiments, but were calculated by Olszewski from the Laplace-Poisson equation for the change of temperature in a gas during adiabatic expansion. According to this equation the temperature of the gas when allowed to expand, as previously described, to a pressure of one atmosphere, is about − 264°. It seems more probable, however, in the light of recent investigations, that the temperature was somewhat above this point. The author also constructed a helium thermometer and compared it with a hydrogen thermometer from the temperature of − 182° to − 210°. The results obtained by the two thermometers agreed very closely.

Liquefaction of Fluorine

In 1895 Dewar remarked that "fluorine is the only widely distributed element in nature which has not been liquefied." Two years later, however, this gas was condensed to the liquid state by Moissan and Dewar.[1] The latter experimenter had previously subjected fluorine to low temperatures with a view of liquefaction, but without success.

[1] *Compl. rend.*, 124, p. 1202; and 125, p. 505, 1897.

The fluorine used by Moissan and Dewar was prepared by the electrolysis of potassium fluoride dissolved in anhydrous hydrofluoric acid. The acid fumes were removed by conducting the gas through a platinum coil, which was surrounded by a mixture of solid carbonic acid and alcohol; after which the fluorine was conducted through platinum tubes filled with perfectly dry sodium fluoride.

The apparatus[1] employed in the liquefaction is represented in figure 42. The glass bulb E is fused to the platinum tube A, which surrounds a smaller platinum tube D. Each of these tubes is provided with a screw-cock so that the communication with the outer air or the current of fluorine can be interrupted at pleasure. In the first series of experiments the glass bulb E was immersed in liquid oxygen, which was contained in a cylindrical vacuum vessel. At the temperature of boiling air (about $-183°$) the fluorine passed through the apparatus without showing any signs of liquefaction. By evaporating the liquid oxygen under reduced pressure, liquid fluorine soon began to collect in the apparatus. The outlet to the fluorine tube was then closed, and the glass bulb soon became filled with a clear yellow, extremely mobile liquid.

---

[1] *Chem. News*, 76, p. 261.

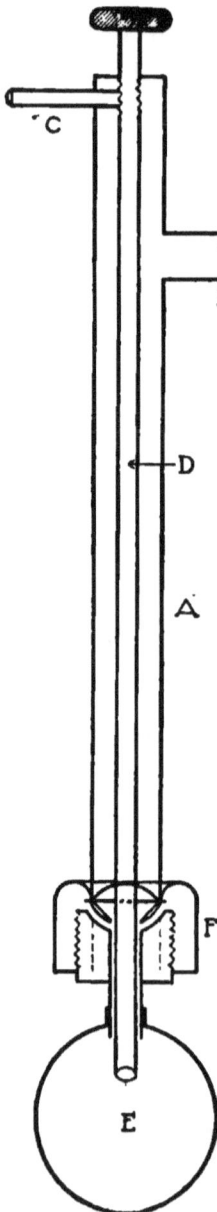

FIG. 42.

At this temperature fluorine did not attack the glass bulb.

The experiment was repeated, using freshly prepared liquid air as the refrigerant. At a temperature of $-190°$ the fluorine condensed to the liquid state. From the two series of experiments the authors calculated the boiling point to be about $-187°$, which is the same temperature as that obtained by Olszewski for the boiling point of argon. The critical temperature was estimated to be about $-120°$, and the critical pressure about 40 atmospheres. The chemical activity of fluorine was greatly reduced at the temperature of boiling oxygen. At this temperature it does not replace iodine, and is without action on phosphorus, boron, silicon, iron, etc. With hydrogen, turpentine, and benzene it reacts with incandescence.

The density of liquid fluorine was determined by placing solid substances of different specific gravities in the liquid. The result of a number of observations gave a density of 1.14. The liquid is soluble in all proportions in liquid oxygen and liquid air.

The authors endeavored to reduce the temperature to the freezing point of fluorine. When the glass bulb was filled to about three-fourths of its capacity, the valves were closed. The liquid oxygen surrounding the glass bulb was then evaporated under a very low pressure, and the temperature sank to − 210°; but even at this low temperature the fluorine retained its characteristic mobility. In a second experiment the liquid fluorine was introduced into a glass tube, which was afterward sealed and kept for some time at a temperature of − 210°, but there were no indications of solidification.

Liquefaction of Hydrogen.

Of the various gases known, hydrogen has presented the most difficult problem to the experimenters on the liquefaction of gases. During the last two decades numerous attempts have been made to liquefy this gas. Reference has already been made to the experiments of Cailletet and Pictet, in which hydrogen was probably obtained in the form of a very fine mist (pp. 119, 135). Wroblewski and Olszewski modified these experiments somewhat by cooling the gas with liquid oxygen, and then allowing the gas to expand (pp. 143, 164). In these observations there can be no doubt as to the formation of a mist of liquid hydrogen. Both of these experimenters endeavored to determine the critical constants and boiling point of this gas.

In 1894 Dewar[1] attacked the problem from a different standpoint. Realizing that, with liquid oxygen and liquid air as refrigerants, the temperature could not be sufficiently reduced for the liquefaction of hydrogen, he endeavored to obtain a liquid, the critical temperature and boiling point of which are considerably lower than those of air and oxygen. This he thought could be accomplished by liquefying a mixture of hydrogen and nitrogen. In regard to the efficiency of this method the author says: "One thing can, however, be proven by the use of the gaseous mixture of hydrogen and nitrogen; viz., that by subjecting it to a high compression at a temperature of − 200°, and expanding the resulting liquid into the air, a much lower temperature than anything that has yet been recorded up to the present time can be reached. This is shown by the fact that

[1] *Chem. News*, 70, p. 115.

such a mixed gas gives, under the conditions, a paste or jelly of solid nitrogen, evidently giving off hydrogen, because the escaping gas burns fiercely. Even when hydrogen containing from two to five per cent of air is similarly treated, the result is a white solid mass (solid air), along with a clear liquid of low density which is so exceedingly volatile that no known device for collecting it has been successful."

During the next year Dewar [1] extended his observations on the liquefaction of hydrogen (p. 198). He says: "Hydrogen, cooled to − 200°, was forced through a fine nozzle under a pressure of 140 atmospheres, and yet no liquid jet could be seen. If, however, hydrogen, previously cooled by a bath of boiling air, is allowed to expand from a pressure of 200 atmospheres over a regenerative coil, a liquid jet can be seen after the circulation has continued for a few minutes, along with a liquid which is in rapid rotation in the lower part of the vacuum-vessel. The liquid did not accumulate, owing to its low specific gravity and the rapid current of gas. These difficulties will doubtless be overcome by the use of a differently shaped vacuum-vessel, and by better isolation."

In May, 1898, Dewar obtained liquid hydrogen

[1] *Proc. Roy. Inst.*, 15, p. 146.

for the first time in a static condition.[1] The method of procedure was similar to that which has just been described. Hydrogen cooled to a temperature of $-205°$, and under a pressure of 180 atmospheres, was allowed to escape continuously from the nozzle of a coil of pipe, at the rate of from ten to fifteen cubic feet per minute, into a doubly silvered vacuum-vessel of special construction. The space surrounding this vessel was kept at a temperature below $-200°$. Soon after the operation was begun, liquid hydrogen began to drop from this vacuum-vessel into a second vessel which was doubly isolated, in that it was surrounded by a third vacuum-vessel. Within five minutes about twenty cubic centimetres of liquid hydrogen had collected in the second bulb. The experiment was interrupted at this point by the solidification of air in the pipes. The liquid obtained was clear and colorless, and showed a well-defined meniscus. No provision was made in this experiment for determining the boiling point of liquid hydrogen, but the author calls attention to an observation which shows that the temperature of liquid hydrogen under atmospheric pressure is extremely low. A long piece of glass tubing, sealed at one end and open at the other, was

---

[1] *Proc. Roy. Soc.*, 63, p. 256, 1898.

cooled by immersing the closed end in liquid hydrogen. The portion of the tube immersed in the liquid was immediately filled with solid air.

A few months later Dewar determined the boiling point and density of liquid hydrogen.[1] The boiling point was measured by means of a platinum resistance thermometer, and was found to be about − 238°. This is a few degrees higher than the values given by Wroblewski and Olszewski. The author thinks that the critical temperature of hydrogen is about − 225°, and the critical pressure about fifteen atmospheres. The density of liquid hydrogen was determined by measuring the volume of gas which is given off when ten cubic centimetres of the liquid are allowed to evaporate. The result of the experiment gave a density of 0.07 for liquid hydrogen, which is only about one-fourteenth that of water.

During the present year Dewar has made a series of observations on the temperature obtained by evaporating liquid hydrogen under reduced pressure.[2] When the liquid was evaporated under a pressure of 25 millimetres of mercury, there were no indications of solidification or loss of mobility. The temperature according to the platinum resistance thermometer was − 239°.1, which

[1] *Chem. News*, 77, pp. 261 and 282, 1898.
[2] *Ibid.*, 79, p. 61, 1899.

Q

is only one degree lower than the boiling point.
The author says that the temperature thus obtained
should have been five or ten degrees below the
boiling point, and adds that the experiment will be
repeated with larger quantities of liquid hydrogen.

Quite recently Dewar has repeated the experiments on the boiling point of hydrogen,[1] and obtained a value somewhat lower than that obtained
in the previous experiments. The author prepared 250 cubic centimetres of liquid hydrogen
for these observations. The temperature of hydrogen, boiling under the atmospheric pressure,
was determined by means of a rhodium-platinum
resistance thermometer and found to be − 246°.
This value is 8° lower than that obtained by
means of the platinum resistance thermometer.

In an addendum to this paper the author calls
attention to some measurements made by means
of a hydrogen thermometer under reduced pressure. This thermometer gave − 182°.5 for the
boiling point of oxygen, and − 252° for the boiling point of hydrogen. If this value is the true
boiling point of hydrogen, it is likely that, by evaporating liquid hydrogen under reduced pressure,
the temperature can be lowered to within ten or
twelve degrees of the absolute zero.

[1] *Chem. News*, March 24, 1899, p. 133.

Liquefaction of Helium

After an elaborate series of experiments with a view of liquefying this gas, Olszewski remarked, "As far as my experiments go, helium remains a permanent gas, and apparently is much more difficult to liquefy than hydrogen" (p. 217). After obtaining liquid hydrogen in a static condition, Dewar [1] placed a sealed glass tube containing helium in the liquid hydrogen. A colorless liquid immediately condensed on the sides of the tube. By placing the same tube in liquid air, boiling under reduced pressure, no condensation was observed. The author concluded that the boiling point of helium is very close to that of hydrogen.

Some Recently Discovered Gases

During the last year Ramsay and Travers have reported three new gases in the atmosphere. The first of these gases,[2] which the authors designated as "krypton," was obtained by evaporating 750 cubic centimetres of liquid air, and collecting the gas from the last ten cubic centimetres. ·After removing the oxygen and nitrogen from this gas, a residue of 26.2 cubic centimetres remained in the vessel. The residual gas had a spectrum dif-

[1] *Proc. Roy. Soc.*, 63, p. 257, 1898.
[2] *Ibid.*, p. 405, 1898.

ferent from that of argon, and was considered by the authors as an elementary substance.

The other two gases, which were called by the authors "the companions of argon," were obtained from liquid argon.[1] The first portions of gas which escape when liquid argon is allowed to evaporate were collected by Ramsay and Travers, and sparked with oxygen gas; after which the excess of oxygen was removed. The residual gas gave a spectrum different from that of argon and krypton, and was called "neon."

During the evaporation of the liquid argon a white solid separated. After the liquid had been completely evaporated, the solid residue was vaporized, and the gas collected. This gas was found to be different from those just described, and was called "metargon."

It is evident from the preceding account that these gases, mixed with air, argon, etc., have all been liquefied. The authors state that "while metargon is a solid at the temperature of boiling air, krypton is probably a liquid, and more volatile at that temperature." Liquid neon is somewhat more volatile than liquid argon.

During this same year Brush published an account of a supposed new gas.[2] The gas was

[1] *Proc. Roy. Soc.*, 63, p. 437, 1898.
[2] Read before the American Association for the Advancement of Science, Aug. 23, 1898.

obtained by exhausting, to a high degree, pulverized soda glass. The resulting gas was found to be a much better conductor of heat than any other known gas. At a pressure of 0.000096 atmospheres the conducting capacity was twenty times that of hydrogen. The author says : " Evidently a new gas of enormous heat-conducting capacity was present, mixed with the last small traces of air." From measurements of the conductivity he concluded that the density of the new gas is only about $\frac{1}{10000}$ that of hydrogen. Owing to the extremely low density the author suggested the name "etherion" for the new gas. In case subsequent observation [1] should confirm the theory of Brush, the new gas will furnish an interesting problem to the experimenters on the liquefaction of gases.

TABLE OF PHYSICAL CONSTANTS

The following table contains the critical constants, boiling points, melting points, etc., of substances which usually occur as gases, and of some of the most common liquids. A few theoretical results are also given for some of the heavier metals. It frequently happens that the results obtained by one experimenter do not agree with those obtained by another. This is especially true

[1] Crookes has already suggested that this gas may be highly exhausted aqueous vapor. —*Chem. News*, 78, p. 221, 1898.

in regard to the critical constants. The names in the right-hand column are given as authority for the critical constants chosen.[1]

Substance.	Critical Temperature	Critical Pressure	Boiling Point	Freezing Point	Color of Liquid	Observer
Acetone . .	+237°.5	60	+ 56°.5	..	Colorless	Sajotschewski
Acetylene . .	+ 37 °.5	68	..	..	Colorless	Ansdell
Air	—	—	—	..		
Alcohol . . .	+243°.6	62.7	+ 78°.3	130°	Colorless	Ramsay and Young
Ammonia . .	+130°	115	— 33°.7	- 75°	Colorless	Dewar
Argon . . .	—121°	50.6	—187°	- 189°.6	Colorless	Olszewski
Arsine . . .	..	..	— 55°	-119°	Colorless	
Carbon dioxide	+ 31°	75	— 78°	— 65°	Colorless	Andrews
Carbon disulphide . . .	+271°.8	74.7	+ 46°	- 110°	Colorless	Sajotschewski
Carbon monoxide . . .	-141°	36	—190°	- 207°	Colorless	Wroblewski
Chlorine . .	+141°	83.9	— 36°.6	- 102°	Yellow	Dewar
Chloroform .	+260°	54	+ 61	— 71°	Colorless	Sajotschewski
Cyanogen . .	+124°	61.7	— 21°	— 34°.4	Colorless	Dewar
Ether . . .	+195° 5	40	+ 35	..	Colorless	Ramsay
Ethylene . .	+ 10° 1	51	—102°.5	—169	Colorless	Dewar
Fluorine . .	- 120°(?)	40 (?)	—187°	..	Yellow	Moissan and Dewar
Gold	+4300°	..	..	+1035°	..	Calculated by Guldberg
Helium . . .	..	..	..	..	Colorless	Dewar
Hydrochloric acid . . .	+51°.25	86	— 35°	—116°	Colorless	Ansdell
Hydrogen . .	..	..	- 252°	..	Colorless	Dewar
Hydrogen sulphide . . .	+100°.2	92	—61°.8	— 85°	Colorless	Dewar
Iron	+ 5200°	..	..	+1500°	..	Calculated by Guldberg
Methane . .	— 95° 5	50	—164°	..	Colorless	Dewar
Nitric oxide .	— 93°.5	71.2	—153°.6	— 167°	Colorless	Olszewski
Nitrogen . .	—146°	35	—194°.5	214°	Colorless	Olszewski
Nitrous oxide.	— 35°.4	75	— 87°.9	115°	Colorless	Dewar
Oxygen . .	—118°	50	—183°	..	Pale blue	Wroblewski
Ozone . . .	..	..	- 125°	..	Indigo blue	
Phosphine . .	..	..	— 85°	—133°	Colorless	
Platinum . .	+ 7000°	..	..	+1800°	..	Calculated by Guldberg
Sulph. dioxide	+155°.4	78.9	— 8°	..	Colorless	Sajotschewski
Water . . .	+358°.1	..	+100°	0°	Colorless	Nadejdine

[1] For further data and literature on critical constants, see Heilborn, *Zeit. Phys. Chem.*, 7, p. 602, 1891.

CONCLUSION

1. *The Three States of Matter*

THE experiments which have been described show that all gases have been condensed to the liquid state. It has also been shown that, with very few exceptions, all gases have been solidified. The results leave no doubt that all of these substances can exist in the gaseous, liquid, or solid state. At the high temperatures which have been obtained by means of the electric furnace, the densest solids have been liquefied and volatilized.

Andrews says the liquid state of matter forms a link between the solid and gaseous states. This link, however, is frequently suppressed, and the solid passes directly into the gaseous condition. Iodine and arsenic are well-known examples of solids which, at the ordinary pressure, sublime directly to the gaseous state without assuming the intermediate liquid condition. Solid carbonic acid behaves in a similar manner. The melting points of these substances are higher than their boiling points. If iodine crystals are placed in a test-tube under sulphuric acid, and the temperature gradually raised, the substance melts and does not vaporize. Prytz[1] has shown that at a pressure of five atmospheres solid carbonic acid does not sublime, but passes directly into the liquid state.

Any substance can exist as a gas at a much lower temperature than that at which it can exist as a liquid. Below the temperature of zero degrees ice slowly sublimes. In the far northern regions the atmosphere always contains aqueous vapor. Pellat[2] has recently shown that iron sublimes very slightly at the ordinary temperature and pressure. Under the proper conditions of temperature and pressure all substances can be made to assume the gaseous, liquid, or solid state. The three states of matter are usually defined as follows: —

1. A gas has neither form nor volume, but tends to expand indefinitely.

2. A liquid has a definite volume, but assumes the form of the vessel in which it is contained.

3. A solid has a definite form and volume.

The relation between the gaseous and liquid states has already been discussed (Chapter III). The change from one condition to the other was found to be gradual and imperceptible. There is

[1] *Phil. Mag.*, 39, p. 308, 1895.
[2] *Zeit. für compr. und flüss., Gase*, 2, p. 95, 1898.

no sharp dividing line. The same is true of the
solid and liquid states. Their properties, in many
cases, are very similar. As the pressure is increased, solids tend more and more to assume the
form of the vessel in which they are contained.
Crystals which have been subjected to enormous
pressure in the crust of the earth are found to
be distorted into various shapes without being
fractured. The crystals seem to flow. The molecules of solids are not rigidly fixed in definite
positions about which they vibrate, but in many
cases move about throughout the entire substance. This has been shown by the experiments of Roberts-Austen on the diffusion of
metals.[1] He showed that, at a temperature of
250°, and even at a temperature of 100°, gold
diffuses throughout the length of solid cylinders
of lead. Similar results were obtained with silver
and gold at a temperature of 800°. The rate
of diffusion in such cases is, of course, very low,
but the process is similar to the diffusion of gases
and liquids.[2]

[1] *Trans. Roy. Soc.*, 187, p. 383, 1896.

[2] Heydweiller has endeavored to find some evidence of critical
phenomena between the liquid and solid states. The change from
the transparent solid to the liquid, he said, appeared to be gradual,
and the melting point increased with the pressure. The experiments were extended to pressures as high as 3500 atmospheres.
Wied. Ann., 64, p. 725, 1898.

2. *Industrial Application of Liquefied Gases*

Liquefied gases have been employed for technical purposes in various directions. The carbonic acid industry is well known. The liquid is used in the preparation of aerated waters, and in the manufacture of salicylic acid. Enormous quantities of carbonic acid are liquefied annually by various establishments. Liquid sulphurous acid is now an ordinary product of commerce. Whenever the gaseous product is desired for laboratory use or technical purposes, it is usually obtained from the liquid. At present, about 4,000,000 kilograms of this liquid are being prepared annually. Liquid acetylene has been introduced for illuminating purposes. Nitrous oxide is now liquefied on a large scale, and used as an anæsthetic for minor surgical operations, especially in dentistry. Liquefied gases are also used in large quantities for the purpose of refrigeration. The ammonia ice-machine is now in operation in most cities. In this process the temperature is lowered and the water frozen by the evaporation of liquid ammonia. Liquid sulphurous acid has also been used for the same purpose. In 1885 Wroblewski said, "Liquid air will be the refrigerant of the future." This prediction, of course, has not yet been fulfilled.

Considerable has been said and written about the use of liquefied gases as a motive power. Numerous attempts have been made to introduce engines or motors for this purpose, but no great success has yet crowned these efforts. The extremely low temperatures which result from the expansion of these liquids to gases at the ordinary pressure are very objectionable in the application of the liquids as a motive power. It is not likely that these liquids will prove of any great service where steam-power is practicable (see p. 211). They may prove to be of considerable value, however, in cases where steam-power is impracticable. It has already been suggested that liquid hydrogen and liquid air may furnish a solution to the balloon problem.

The industry of liquefied gases is growing rapidly. Every year sees a wider application of these liquids. A complete discussion of this subject, however, does not fall within the scope of a work of this nature.[1]

3. *Physiological Action at Low Temperatures*

Some interesting observations have been made in regard to the influence of very low temperatures on living organisms. In 1870 Cohn[2] made use of

---

[1] For further discussion of the subject, see references on p. 183.
[2] Cohn's *Beiträge zur Biologie der Pflanzen*, 1870, 2, p. 221.

freezing mixtures, and subjected bacteria, for a period of 12 hours, to temperatures varying from 0° to — 18° without destroying their activity. Melsens[1] used solid carbonic acid and exposed yeast and vaccine lymph to a temperature of —78° without destroying the life of the organisms. Pictet and Yung[2] subjected various bacteria to low temperatures. They reduced the temperature by the evaporation of liquid sulphurous and carbonic acids. The organisms were subjected for 20 hours to a temperature of —70°; for 89 hours to a temperature of —76°; and finally for 20 hours to a temperature of —130° (= —202 F.). Yeast ferment showed no alterations under the microscope, but lost its power of fermentation. *Bacillus anthracis* and several other micro-organisms retained their virulence when injected into living animals.

In 1885 a very elaborate series of experiments were made by Coleman and McKendrick[3] in regard to the effect of low temperatures on certain bacteria. Thirty samples of fresh meat were placed in two-ounce white glass phials. The bottles were then carefully closed with corks which had been steeped in mastic varnish, and the

[1] *Compt. rend.*, 70, p. 629; and 71, p. 325.
[2] *Ibid.*, 98, p. 747, 1884.
[3] *Proc. Roy. Inst.*, 11, p. 309.

necks of the corked bottles were immersed in mol-
ten sealing wax. The specimens were treated as
follows : —

6 samples were exposed to a temperature of $-17°$ for 65 hours.
6   "   "   "   "   "   " -29 "   "
6   "   "   "   "   "   " -34 "   "
6   "   "   "   "   "   " -40 "   "
6   "   "   "   "   "   " -62 "   "

Within ten or twelve hours after removal to a
warm room, signs of putrefaction were visible in
all of the bottles, and in the course of a few days
the putrefactive process was fully established.
Other samples were then exposed to a temperature
of $-83°$ for a period of 100 hours with similar
results. Samples of fresh milk, which had been
hermetically sealed and subjected to a temperature
of $-62°$ for eight hours, curdled when kept in a
warm room.

The following observations were made by subjecting a rabbit to low temperatures : —

	Temperature	Pulse	Respiration
Before the experiment	99°.2 F.	160 per min.	45 per min.
After 30 min. exposure to $-93°$	94.2 "	—	—
After 60 min. exposure to $-100°$ . . .	43.2 "	40 per min.	Scarcely Perceptible

In this condition reflex action became almost imperceptible. When placed in a warm room the animal completely recovered.

The effect of low temperatures on cold-blooded animals is entirely different from that on warm-blooded animals. The author states that, at low temperatures, a frog became as hard as a stone in from ten to twenty minutes, while the warm-blooded animal produced within itself sufficient heat to enable it to remain soft and comparatively warm during an exposure for one hour to a temperature of $-$ 100° F. The production of heat, however, was not equal to the loss, and the animal was continually losing ground; the bodily temperature having decreased 56° during the exposure.

In 1893 Pictet[1] made a second series of experiments. He describes the struggle of nature against external attacks as follows : —

When a dog is placed in a copper receiver which is cooled down to from $-$ 60° to $-$ 90°, its temperature rises about one-half degree during the first ten minutes. After ninety minutes the temperature falls one degree. Then follows a point where the struggle is given up; the temperature falls rapidly, and the animal dies suddenly. The author made similar experiments by exposing his arm to

[1] *Chemiker Zeitung*, 1893, p. 1337; *Chem. News*, 68, p. 312.

low temperatures. The only pain occurs within the arm on the periosteum, the epidermis experiencing no pain whatever.

All insects resist a temperature of − 28°, but not − 35°. Myriapods resist a temperature of − 50°, and snails − 120°. The eggs of birds lose their vitality at − 2° or − 3°, the eggs of ants at 0°, and the eggs of the silkworm at − 40°. Infusoria died at − 90°, and bacteria retained their virulence after an exposure to a temperature of − 213°. The temperature in the case of the bacteria was lowered by means of frozen atmospheric air.

In 1894 Pictet[1] continued his observations, and exposed himself to a temperature of − 110°. The body was well protected by clothing during this exposure, and the legs were kept in motion. Pictet says that, for a number of years previous to this exposure, he had been suffering from indigestion, and then adds that the exposure to this low temperature effected almost a complete cure. He suggests that the influence of low temperatures on physiological action may prove to be of great therapeutic value.

Dewar[2] states that McKendrick tried the effect of low temperatures on the spores of micro-organ-

---

[1] *Compt. rend.*, 119, p. 1016.
[2] *Chem. News*, 67, p. 211, 1893.

isms. Samples of flesh, blood, milk, and similar substances were sealed in glass tubes and exposed for a period of one hour to a temperature of − 182°. On standing at the ordinary temperature for several days these substances became putrid. Seeds of different kinds germinated after being exposed to a temperature of − 182°. Dewar considers the experiments with seeds as an evidence in favor of the possibility of Lord Kelvin's suggestion, that life may have been brought to the newly cooled earth upon a seed-bearing meteorite.

4. *Properties of Matter at Low Temperatures*

At the low temperatures obtained by the evaporation of liquid air, etc., the properties of most substances are materially modified. All organic compounds and, with very few exceptions, all inorganic compounds, are solids at that temperature. Rubber placed in liquid air loses its elasticity and becomes as brittle as glass, but regains its original condition when the temperature is allowed to rise. Tin and many other metals also become very brittle. Kreutz[1] has shown that the power of absorbing light is considerably modified at low temperatures. Many substances (mercuric

[1] *Phil. Mag.* [5], 39, p. 209.

iodide, mercuric oxide, etc.) change color under such conditions.

Pictet[1] and Dewar[2] have investigated the influence of low temperatures on the phenomenon of phosphorescence. Many substances lose their power of phosphorescence at low temperatures, while other substances, which exhibit only a feeble phosphorescence at the ordinary temperature, become more phosphorescent. A temperature of $-80°$ is sufficient to stop all sensible emission from previously excited calcium sulphide, but it does not prevent the unexcited calcium sulphide from absorbing light-energy which can be evolved at higher temperatures.

Zakrzewski[3] has shown that the specific heat of silver changes about three per cent in the interval from $0°$ to $-100°$. Important results have also been obtained by Dewar and his associates on the resistance of metals at low temperatures.[4] The resistance of any metal to the passage of the electric current decreases with decreasing temperature. This law, however, does not hold true for alloys. Dewar says: "The results point to the conclusion that pure metals have no resistance near the abso-

---

[1] *Compt. rend.*, Sept. 24, 1894.
[2] *Proc. Roy. Inst.*, 14, p. 665, 1895.
[3] *Phil. Mag.* [5], 39, p. 191.
[4] See references on p. 178.

R

lute zero of temperature. With alloys there is
little change in resistance. In the case of carbon
the resistance decreases with increasing temperature. At the temperature of the electric arc, carbon appears to have no resistance."

Some interesting observations have been made
on chemical action at low temperatures. Dewar
says : " At a temperature of − 200° the molecules
of matter seem to be drawing near to what might
be called the 'death of matter' so far as chemical
action is concerned." At this temperature yellow
phosphorus and liquid oxygen show no signs of
reaction. In 1861 Loir and Drion[1] showed that
liquid ammonia in contact with concentrated sulphuric acid does not react at first. Most of the
more recent experimenters on the liquefaction of
gases have made observations on chemical action
at low temperatures. The references in most cases
have already been given. In general, chemical
action ceases under these conditions. Metallic
sodium and potassium can be thrown into liquid
oxygen without action. Pictet[2] has shown that
sodium does not react with aqueous hydrochloric
acid (15 % solution) at a temperature of − 80°.

In some cases, however, chemical action takes
place at very low temperatures. Liquid ethylene

[1] *Phil. Mag.* [4], 20, p. 202.
[2] *Compt. rend.*, 115, p. 814, 1894.

reacts with chlorine and bromine at a temperature of $-102°.5$. Dewar[1] has shown that photographic action takes place at a temperature of $-180°$. The photographic action seems to be reduced by about 80 per cent. The author states that "it is certain that the Eastman film is fairly sensitive to photographic action at a temperature of $-200°$." Quite recently it has been shown that liquid fluorine at a temperature of $-187°$ reacts, with evolution of light and heat, with hydrogen, benzene, turpentine, etc.

The preceding observations merely indicate the various directions in which liquefied gases have been employed. More than a century ago the great Lavoiser predicted that, were the earth suddenly placed in a very cold region, the atmosphere would cease to exist as an invisible fluid, but would return to the liquid state, and new liquids, of which we have no knowledge, would be produced. This prediction has been thoroughly confirmed. One by one the various gases have been condensed to the liquid state, and the term "permanent gas" has lost its significance. Along with the development of the methods employed in the liquefaction of gases, new industries of great commercial value have been opened up. The extremely low tem-

[1] *Proc. Roy. Inst.*, 14, p. 665.

peratures which can be obtained by means of these liquids have broadened the range of scientific research. Numerous and important observations in this direction have already been made, yet the investigations at low temperatures are only in their infancy. For many years the scientific world has been speculating in regard to the probable condition and properties of matter at the absolute zero of temperature, — a temperature which experimenters have sought in vain to reach. Every year, however, shortens the distance to travel, and at present only a few degrees separate us from the desired goal.

INDEX TO AUTHORS

INDEX TO SUBJECTS

Electric conductivity of metals at low temperatures, 178.
Ethane, attempt to solidify, 159.
Ether, vaporization of, in closed tubes, 18, 19, 20, 81.
thermal transparency of, 175.
Etherion, 229.
Ethylene, liquefaction of, 48.
solidification of, 158.
thermal transparency of liquid, 175.
Euchlorine, liquefaction of, 26, 51.
solidification of, 51.
Expansion of gases, influence of temperature on, 12.
Explosible ether, 37.

Fluoboron, liquefaction of, 50.
Fluorescence, 90.
Fluorine, liquefaction of, 218-221.
attempt to solidify, 221.
reaction of, at low temperatures, 220, 221, 243.
Fluosilicon, liquefaction of, 49.

Gas, definition of, 6, 81, 95, 232.
kinetic theory of, 102-104.
relation to liquid, 84-112.
Gaseous mixtures, critical constants of, 82.

Helium, attempts to liquefy, 216-218.
liquefaction of, 227.
Hydriodic acid, liquefaction of, 49.
solidification of, 49.
Hydrobromic acid, liquefaction of, 49.
solidification of, 49.
Hydrocarbons, liquefaction of, 86, 187.
Hydrochloric acid, liquefaction of, 15, 17, 23, 27, 81.
solidification of, 158.
Hydrochloric ether, expansion of liquid, 63.

Hydrofluoric acid, solidification of, 158.
Hydrogen, attempts to liquefy, 40, 43, 54, 60, 71, 114, 119, 135.
boiling point of, 164, 225, 226.
density of liquid, 225.
liquefaction of, 143, 153, 221-226.
Hydrogen selenide, solidification of, 158.
Hydrogen sulphide, see Sulphuretted Hydrogen.

Ice machines, 181-183, 234.
Ice, sublimation of, 232.
Industrial application of liquefied gases, 234, 235.
Iodine, solubility of, in liquid carbonic acid, 91, 92.
liquefaction of, 232.
Iron, sublimation of, 232.

Krypton, 227, 228.

Liquid, definition of, 95, 232.

Marsh gas, see Methane.
Matter, properties of, at low temperatures, 240-244.
Mechanical equivalent of heat, 205.
Melting point, table of, 230.
Metargon, 228.
Methane, liquefaction of, 114, 145, 159.
density of liquid, 159.
solidification of, 158.
Muriatic acid, see Hydrochloric Acid.

Naphtha, vaporization of, in closed tubes, 18.
Neon, 228.
Nitric oxide, attempts to liquefy, 43, 54, 62, 71.
liquefaction of, 116.
solidification of, 158.

www.ingramcontent.com/pod-product-compliance
Lightning Source LLC
Chambersburg PA
CBHW021522210326
41599CB00012B/1344